AF598073

Nanometer Variation-Tolerant SRAM

Mohamed H. Abu-Rahma
Mohab Anis

Nanometer Variation-Tolerant SRAM

Circuits and Statistical Design for Yield

Mohamed H. Abu-Rahma
Qualcomm Incorporated
San Diego, CA
USA

Mohab Anis
Electronics Engineering Department
School of Sciences and Engineering
The American University in Cairo
New Cairo
Egypt

ISBN 978-1-4614-1748-4 ISBN 978-1-4614-1749-1 (eBook)
DOI 10.1007/978-1-4614-1749-1
Springer New York Heidelberg Dordrecht London

Library of Congress Control Number: 2012943374

Printed on acid-free paper

Springer is part of Springer Science+Business Media (www.springer.com)

To my mother, Eman
my father, Hassan
my wife, Enas
and my beautiful daughter, Mariam

Mohamed

To Soheir Elnahas and Hussein Anis

Mohab

Preface

Aggressive scaling of CMOS technology in sub-90 nm nodes has created huge challenges for SRAM design. Variations due to fundamental physical limits, subwavelength lithography, and device aging are increasing significantly with technology scaling. In addition, SRAMs typically use the smallest devices in a technology, which make them suffer from the largest variations. Moreover, in recent technologies, there is a high demand to integrate large embedded memories in microprocessors and SoCs, and memory yield has huge impact on the chip cost.

The increase of variations with technology scaling decreases the robustness of SRAM, as a result, memory yield decreases, and the memory minimum supply voltage (V_{min}) is becoming the limiting factor of voltage scaling in newer technologies. In this book, we offer a comprehensive overview of variation-tolerant techniques used to mitigate the negative impacts of variations, and improve memory robustness. The topic spans a wide area between device technology, custom circuits, statistical CAD, and architecture. This book presents state-of-the-art research in the areas of variability, variation-tolerant SRAM circuits, and assist techniques, statistical design, and SRAM silicon characterization. This book is intended for graduate students, researchers as well as practicing engineers interested in understanding the key challenges facing SRAM design in nanometer technologies, and the techniques used to address them.

San Diego, CA, USA, May 2012 — Mohamed H. Abu-Rahma
Cairo, Egypt, May 2012 — Mohab Anis

Acknowledgments

First and foremost, all praise is due to Allah, the almighty, for giving us the blessings and strength to complete this work.

We would like to thank Qualcomm's staff for their support, in particular, we would like to express our gratitude to Sei Seung Yoon and Esin Terzioglu for their technical leadership and support. Mohamed would like to thank his teammates at Qualcomm, especially, Muhammad Nasir, Dongkyu Park, ChangHo Jung, and all the members of the Memory Design Group for many valuable discussions. Also, he deeply thanks his friends Hassan Hassan, Hany Bakir, and Mostafa Emam for their help and encouragement.

We owe a large debt of gratitude to our professors from whom we have learned invaluable lessons about research. Mohamed would like to specially thank Prof. Wael Fikry, Prof. Mohamed Dessouky and Prof. Hany Fikry from Faculty of Engineering, Ain Shams University. Many thanks are due to the editorial staff of Springer including Charles B. Glaser and Elizabeth Dougherty, who were very supportive and encouraging. We would also like to thank Jessica Moore for her professional help in editing the book.

Writing this book would not have been possible without the encouragement, dedication, and love of our families. Our deepest gratitude goes to our parents for their never-ending support, and for always remembering us in their prayers. No words of appreciation could ever reward them for everything they have done for us. We are, and will ever be, indebted to them for all the achievements in our lives. Mohamed is deeply thankful to his sisters for their support. Throughout the course of writing this book, our wives provided endless support, encouragement, patience, and love. We thank them for all the sacrifices they have made.

Contents

Acronyms

ATE	Automatic/automated test equipment
BER	Bit error rate
BISR	Built-in self-repair
BIST	Built-in self-test
BTBT	Band to band tunneling
CAD	Computer Aided design
CD	Critical dimension
CDF	Cumulative distribution function
CLSA	Current latch sense amplifier
CMOS	Complementary metal oxide semiconductor
D2D	Die-to-die variations
DAC	Digital to analog converter
DIBL	Drain induced barrier lowering
DRV	Data retention voltage
ECC	Error correction codes (or circuits)
EUV	Extreme ultraviolet lithography
EVT	Extreme value theory
FBB	Forward body bias
FET	Field effect transistor
FinFET	Fin field effect transistor (3D transistor)
GEV	Generalized extreme value distribution
GIDL	Gate induced drain leakage
IC	Integrated circuits
IS	Importance sampling
LER	Line edge roughness
MC	Monte Carlo
MCU	Multiple cell upsets
MPFP	Most probable failure point
NBTI	Negative bias temperature instability
OPC	Optical proximity correction
OPE	Optical proximity effects

PBTI	Positive bias temperature instability
PDF	Probability density function
PVT	Process voltage temperature
RBB	Reverse body bias
RDF	Random dopant fluctuation
RTN	Random telegraph noise
RTS	Random telegraph signals
RF	Register file
SA	Sense amplifier
SCE	Short-channel effects
SER	Soft error rate
SEU	Single event upset
SNM	Static-noise Margin
SoC	System-on-a-chip
SOI	Silicon-on-insulator technology
SRAM	Static random access memory
SSTA	Statistical static timing analysis
STA	Static timing analysis
VLSI	Very large scale integration
VTC	Voltage transfer characteristics
WID	Within-die variations
WM	Write margin

Chapter 1
Introduction

1.1 Motivation: Variation-Tolerant SRAM Design

Four decades of technology scaling in CMOS has been the largest driver of growth for the electronics industry. Scaling of CMOS transistors has led to chips with more than one billion transistors in modern ICs and a wide range of products with very high levels of integration [1]. However, the aggressive scaling of CMOS technology in sub-90 nm nodes has created huge design challenges. Due to process control limitations, manufacturing tolerances in process technology are not scaling at the same pace as transistor channel length [2–6]. Moreover, variations due to fundamental physical limits are increasing significantly with technology scaling [2, 7, 8]. Due to all these sources, statistical parameter variations worsen with successive technology generations, and variability is currently one of the biggest challenges facing the semiconductor industry [5]. This variability has been affecting analog design for some time, and now it is dramatically impacting static random access memory (SRAM) design at nanometer technology nodes.

Process variations have a strong impact on SRAM because they increase bitcell failure probability especially at low voltage [9–11]. With the exponential increase in embedded SRAM content in microprocessors and system on a chip (SoCs), SRAM yield has a strong impact on the overall product yield (and of course cost) [5, 12]. In addition, SRAM uses the most aggressive design rules to achieve the highest possible integration density, which makes SRAM the most sensitive circuit for process variations. Due to the ubiquitous nature of embedded memories, SRAM yield loss due to variability is likely the dominant cause of yield loss in modern ICs. Therefore, it is not surprising that the main focus of SRAM design in sub-90nm technologies is the development of variation-tolerant techniques to reduce SRAM sensitivity to variations and increase memory yield [9–11].

In order to improve SRAM design in the nanometer regime, it is critical to explore variation-tolerant design techniques that involve circuits, architectures, and utilize advanced statistical CAD. This book addresses the increase of variability in nanometer technologies and focuses on SRAM variation-tolerant design, and covers

M. H. Abu-Rahma and M. Anis, *Nanometer Variation-Tolerant SRAM*,
DOI: 10.1007/978-1-4614-1749-1_1, © Springer Science+Business Media New York 2013

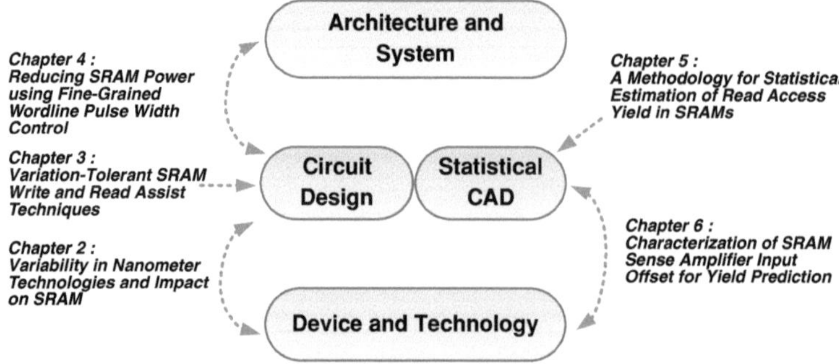

Fig. 1.1 Different levels of design abstraction studied in this book

state-of-the-art circuit techniques and statistical CAD to optimize SRAM performance and yield in nanometer technologies. This book intends to show the importance of addressing variability at different levels of design abstractions, as shown in Fig. 1.1. Investigating multiple levels of abstraction maximizes the effectiveness of variation-tolerant designs helps in bridging the gap between devices, circuits and CAD, and architectures.

1.2 Book Outline

To set the stage for our discussion on variation-tolerant design, we begin in Chap. 2 by looking at SRAM scaling trends and reviewing the sources of variability in nanometer technologies, including process, device noise, reliability, and environmental variations. We also examine how variations, soft errors, and hard defects affect SRAM operation, and discuss the impact of variations on the minimum operation supply voltage $V_{\min}$. Due to the similarities between SRAM and logic circuits, we briefly discuss how variations affect logic circuits and the techniques used to mitigate variability in logic circuits.

In Chap. 3, we focus on circuit techniques to mitigate SRAM failures due to process variations. We start by discussing different stability metrics used to assess SRAM stability margins. We then explain circuit techniques involving bitcell design to enhance stability margins for low voltage operation. Next, we present state-of-the-art write assist and read assist techniques to improve the SRAM design margins. As as case study, we implement a single supply read assist in a 512 kb memory in 45 nm technology.

In Chap. 4, variation-tolerant architectures are briefly discussed. We show that in the presence of large variations, SRAM power consumption increases to ensure correct read operation. A variation-tolerant architecture is proposed to reduce SRAM power consumption using a closed loop system that controls the wordline pulse width.

Statistical simulation is used to estimate power savings enabled by the proposed system in industrial 45 nm technology.

In Chap. 5, we discuss statistical CAD techniques used in SRAM design. Conventional (i.e., Monte Carlo, sensitivity analysis) as well as state-of-the-art techniques (i.e., importance sampling, most probable failure point, and statistical blockade) are presented. As a case study of statistical CAD, we present a methodology for statistical estimation of read access yield. The proposed yield estimation flow provides yield and performance tradeoffs during design, which can be used to optimize the memory performance and architecture. Results from this methodology are verified with measured SRAM yield from 45 nm technology.

In Chap. 6, we discuss the importance of silicon characterization to improve SRAM yield. A process control monitor to measure sense amplifier offset is proposed and implemented in 28 nm technology. The monitor accurately measures SA offset from a large sample size and accounts for all layout proximity effects. The all-digital design of the monitor makes it appropriate for low voltage testing, high speed data collection, and ease of migration to newer technologies. Statistical yield estimation using the sense amplifier offset correlates well with measured yield for a 512 kb SRAM. The monitor improves the accuracy of SRAM silicon yield validation, which is becoming increasingly important with technology scaling and the resulting increase in random variations.

References

1. T.-C. Chen, Where is CMOS going: trendy hype versus real technology, in *Proceedings of the International Solid-State Circuits Conference ISSCC*, 2006, pp. 22–28
2. S.R. Nassif, Modeling and analysis of manufacturing variations, in *Proceedings of IEEE Custom Integrated Circuits Conference*, 2001, pp. 223–228
3. H. Masuda, S. Ohkawa, A. Kurokawa, M. Aoki, Challenge: variability characterization and modeling for 65- to 90-nm processes, in *Proceedings of IEEE Custom Integrated Circuits Conference*, 2005, pp. 593–599
4. B. Wong, A. Mittal, Y. Cao, G.W. Starr, *Nano-CMOS Circuit and Physical Design* (Wiley-Interscience, New York, 2004)
5. The International Technology Roadmap for Semiconductors (ITRS), http://public.itrs.net
6. J. Tschanz, K. Bowman, V. De, Variation-tolerant circuits: circuit solutions and techniques, in *DAC '05: Proceedings of the 42nd Annual Conference on Design Automation*, 2005, pp. 762–763
7. D. Frank, R. Dennard, E. Nowak, P. Solomon, Y. Taur, H.S. Wong, Device scaling limits of Si MOSFETs and their application dependencies. Proc. IEEE **89**(3), 259–288 (2001)
8. J.A. Croon, W. Sansen, H.E. Maes, *Matching Properties of Deep Sub-Micron MOS Transistors* (Springer, New York, 2005)
9. K. Agarwal, S. Nassif, Statistical analysis of SRAM cell stability, in *DAC '06: Proceedings of the 43rd Annual Conference on Design Automation*, 2006, pp. 57–62
10. S. Mukhopadhyay, H. Mahmoodi, K. Roy, Statistical design and optimization of SRAM cell for yield enhancement, in *Proceedings of International Conference on, Computer Aided Design*, 2004, pp. 10–13
11. R. Heald, P. Wang, Variability in sub-100nm SRAM designs, in *Proceedings of International Conference on, Computer Aided Design*, 2004, pp. 347–352
12. Y. Zorian, Embedded memory test and repair: infrastructure IP for SOC yield, in *Proceedings the International Test Conference (ITC)*, 2002, pp. 340–349

Chapter 2
Variability in Nanometer Technologies and Impact on SRAM

2.1 SRAM Scaling Trends

In today's SoCs and microprocessors, embedded SRAM comprises a large portion of chip area. Figure 2.1 shows an example of a modern microprocessor where embedded SRAM (caches) consumes significant chip area. As shown in Fig. 2.2, SRAM area is expected to exceed 90 % of overall chip area by 2014 [1] because of the demand for higher performance (multiprocessing and multicores), lower power, and higher integration. The large contribution of SRAM has strong impact on chip cost and yield.

To increase memory density, memory bitcells are scaled to reduce their area by 50 % each technology node, as shown in Fig. 2.3. High density SRAM bitcells use the smallest devices in a technology, making SRAM more vulnerable for variations [3, 4]. For example, in state of the art 28 nm technology, a high density bitcell area is approximately $0.12\,\mu m^2$, as shown in Fig. 2.4. This compact bitcell enables an integration of $7.9\,\text{Mbit/mm}^2$.

While process variation degrades performance and increases leakage in random logic, its impact on SRAM is much stronger. In advanced CMOS technology nodes, the predominant yield loss comes from the increase in process variations, which strongly impacts SRAM functionality as the supply voltage is reduced [5–9]. In particular, local random variations due to and line edge roughness (LER) strongly decrease the robustness of SRAM operation. Figure 2.5 shows that V_{th} variation for SRAM devices increases significantly with scaling, which poses a major challenge for SRAM design [10]. In the following sections we look at the different sources of variations that affect SRAM.

M. H. Abu-Rahma and M. Anis, *Nanometer Variation-Tolerant SRAM*,
DOI: 10.1007/978-1-4614-1749-1_2,

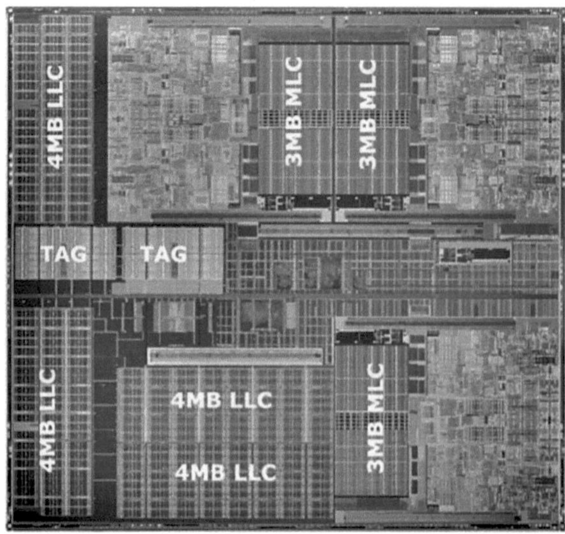

Fig. 2.1 Chip micrograph for a modern microprocessor illustrating the large contribution of embedded SRAM (caches) in the total chip area [2]

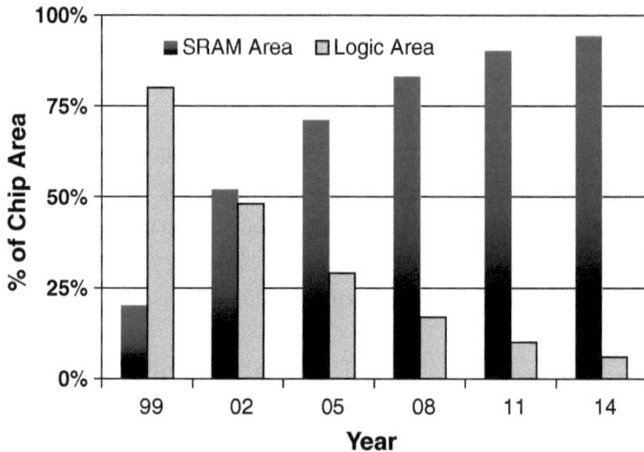

Fig. 2.2 SRAM and logic area versus technology scaling. SRAM dominates chip area in modern SoCs and microprocessors [1]

2.2 Classification of Sources of Variation

Variation is the deviation from intended values for structure or a parameter of concern. The electrical performance of modern IC is subject to different sources of variations that affect both the device (transistor) and the interconnects. For the purposes of circuit design, the sources of variation can broadly be categorized into two classes [12–15]:

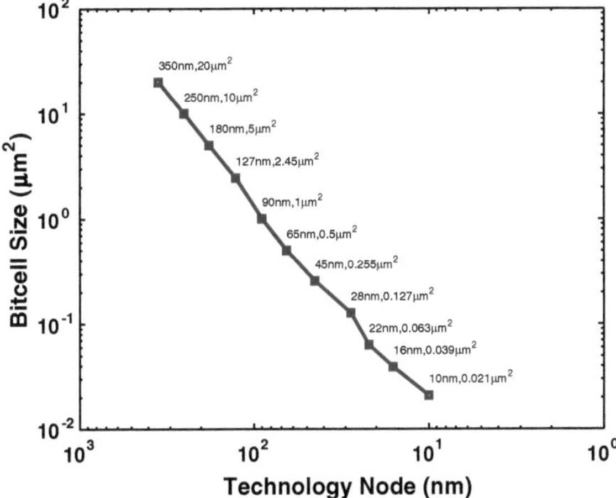

Fig. 2.3 SRAM bitcell area scaling from 350 nm down to 10 nm technology nodes. Bitcell area continues to scale by 50 % for each node

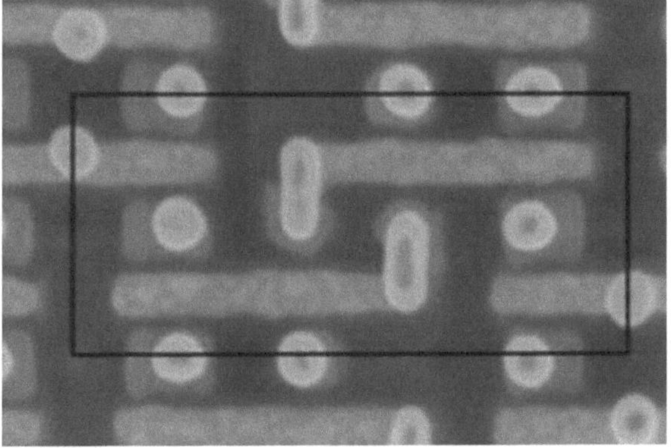

Fig. 2.4 SEM for a 0.12 μm^2 6T bitcell in 28 nm technology node [11]

- **Die-to-Die (D2D)**: also called global or inter-die variations affect all devices on the same chip in the same way (e.g., they may cause all the transistors' gate lengths to be larger than a nominal value).
- **Within-Die (WID)**: also called local or intra-die variations, correspond to variability within a single chip, and may affect different devices differently on the same chip (e.g., devices in close proximity may have different V_{th} than the rest of the devices).

D2D variations have been a longstanding design issue, and are typically accounted for during circuit design with using corner models [12, 13, 16]. These corners are chosen to account for the circuit behavior under with the worst possible variation, and

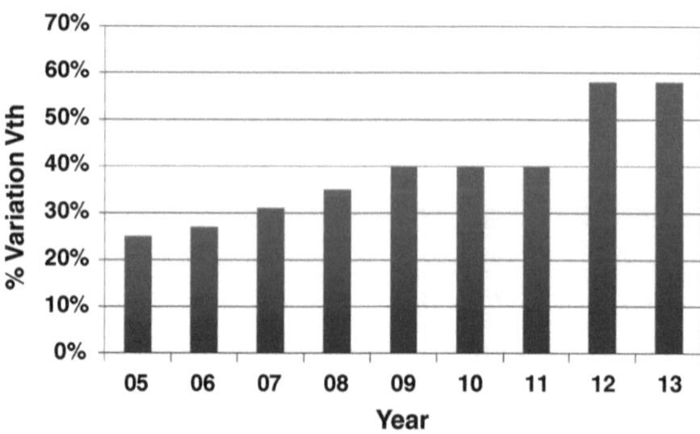

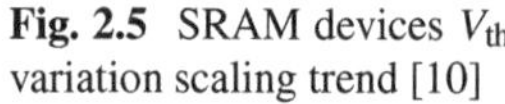
Fig. 2.5 SRAM devices V_{th} variation scaling trend [10]

were considered efficient in older technologies where the major sources of variation were D2D variations.

However, in nanometer technologies, WID variations have become significant and can no longer be ignored [17–23]. As a result, process corners-based design methodologies, where verification is performed at a small number of design corners, are currently insufficient.

WID variations can be subdivided into two classes [12–15]:

- **Random variations**: as the name implies, are sources that show random behavior, and can be characterized using their statistical distribution.
- **Systematic variations**: show certain variational trends across a chip and are caused by physical phenomena during manufacturing such as distortions in lenses and other elements of lithographic systems. Due to difficulties in modeling this type of variation, they are usually modeled as random variations with certain value of spatial correlation.

Other classifications for variability include time-dependency (long or short), static, dynamic, device, interconnect, and environment. In the following sections, we present an overview of the device, interconnect, and environment sources of variations.

2.3 Device Variability

Process variations impact device structure and therefore change the electrical properties of the circuit. In the following subsections, we review the main sources of variations that affect device performance.

2.3.1 Random Dopant Fluctuations

As CMOS devices are scaled down, the number of dopant atoms in the depletion region decreases, especially for a minimum geometry device. Due to the discreteness of atoms, there is statistical random fluctuation of the number of dopants within a

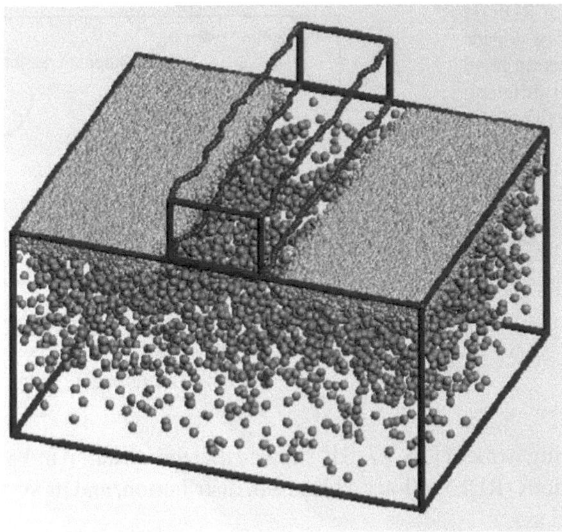

Fig. 2.6 Atomistic process simulation incorporating random dopant fluctuation (RDF) and line edge roughness (LER) as the sources of intrinsic fluctuation [30]. The *dots* show the dopant atoms that determine the device's threshold voltage.

given volume [24–29]. This fluctuation in the number of dopants in the transistor channel results in variations in the observed threshold voltage V_{th} for the device. Figure 2.6 shows how dopants are placed in the transistor channel.

For example, in a uniformly doped $W = L = 0.1\,\mu\text{m}$ NMOS, if the doping concentration is $N_a = 10^{18}\,\text{cm}^{-3}$ and depletion width at zero body bias is $W_{\text{dmo}} = 350\,\text{Å}$, the average number of acceptor atoms in the depletion region can be calculated as $N = N_a.L.W_{\text{dmo}} = 350$ atoms. Due to the statistical nature of dopants, the actual number fluctuates from device to device with a standard deviation following a Poisson's distribution, and therefore $\sigma_N = \langle(\Delta N)^2\rangle^{1/2} = \sqrt{N}$, which for our example yields $\sigma_N = 18.7$, a significant fraction of the average number N (σ_N/N is 5 % in this example). Variation in the number of dopant atoms directly affects the threshold voltage of a MOSFET, since V_{th} depends on the charge of the ionized dopants in the depletion region [25].

These fluctuations were anticipated long ago [26, 31] and have always been important for SRAM bitcells and analog circuits, due to their sensitivity to mismatch [26, 31]. With technology scaling, the number of dopants in the depletion region has been decreasing steadily, as shown in Fig. 2.7. The decrease has been roughly proportional to L, so that we are now into the regime in which the smallest FETs have less than few hundred dopants determining the threshold voltage [32]. Following Poisson statistics, fluctuations in the dopant number have a standard deviation equal to the square root of the number of dopants, which causes a large increase in V_{th} variation as shown in Fig. 2.7.

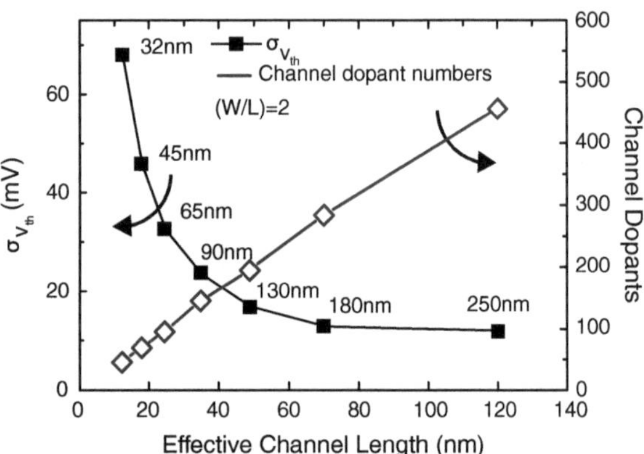

Fig. 2.7 Impact of RDF on σV_{th} and number of dopant atoms in the depletion layer of a MOSFET for different technology nodes [32, 33]

The pioneering work of [26, 27, 31] showed that the variation in V_{th} due to random dopant fluctuations (RDF) follows a Gaussian distribution, and its standard deviation can be modeled as:

$$\sigma_{V_{th}} = \left(\sqrt[4]{2q^3 \varepsilon_{Si} N_a \phi_B}\right) \times \frac{T_{ox}}{\varepsilon_{ox}} \times \frac{1}{\sqrt{3\,W\,L}} \tag{2.1}$$

where q is the electron charge, ε_{Si} and ε_{ox} are the permittivity of the silicon and gate oxide, respectively, N_a is the channel dopant concentration, ϕ_B is the difference between Fermi level and intrinsic level, T_{ox} is the gate oxide thickness, and W and L are the channel width and channel length for the transistor, respectively.

Equation (2.1) shows that $\sigma_{V_{th}}$ is inversely proportional to the square root of the active device area. Hence, the transistors can be sized up to mitigate variations, which is one of the main techniques used in analog design to reduce mismatch between transistors [34]. Moreover, V_{th} variation is largest in SRAM devices, which typically use the smallest sizes in a technology. In addition, Eq. (2.1) shows that variation increases with technology scaling. Figure 2.5 shows the large increase in $\sigma_{V_{th}}$ with technology scaling for SRAM devices. Relative variation can reach about 50 % of V_{th} in advanced technologies which has strong impact on SRAM operation.

2.3.2 Line Edge Roughness

Gate patterning introduces a non-ideal gate edge; this imperfection is referred to as LER, as shown in Fig. 2.8. As device scaling continues into sub-50 nm regime, LER is expected to become a significant source of variation due to its direct impact on $\sigma_{V_{th}}$ [24, 35, 36].

Figure 2.9 shows that nanometer technologies use light sources with wavelengths which are much larger than the minimum feature size [17] which increases gate variation due to LER. In addition, the patterning of features smaller than the wave-

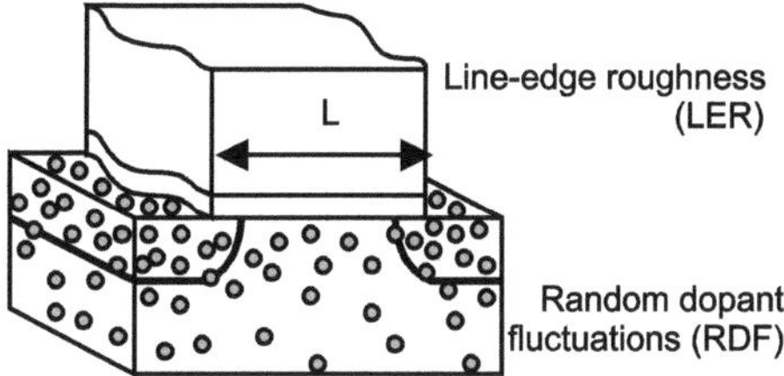

Fig. 2.8 Primary sources of variation in nanometer technologies: RDF and LER [33]

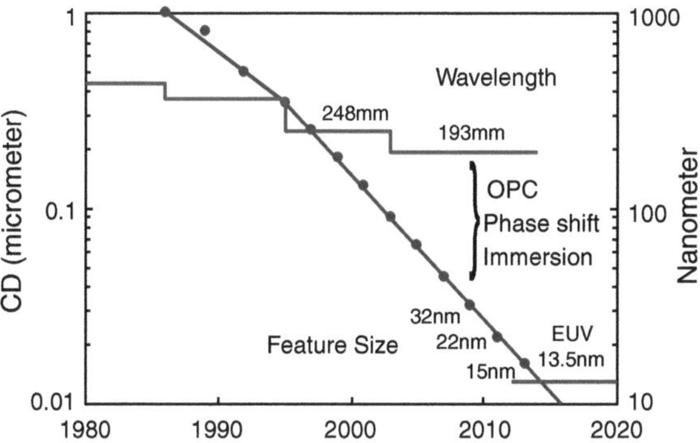

Fig. 2.9 Lithography wavelength scaling for different technology nodes critical dimension (CD). Recent technology nodes used 193 nm optical lithography with enhancements such as optical proximity correction (OPC), aperture improvement, and immersion technology to extend the lithography capabilities down to 15 nm [36]

length of light used in lithography results in distortions due to the diffraction of light, referred to as optical proximity effects (OPE) [12, 36, 37]. OPEs cause large variations in defining the minimum feature sizes (e.g., critical dimension CD), and increases LER variations [15, 36, 38].

Controlling LER variations is extremely difficult since the variations do not scale with technology; in other words, the improvements in the lithography process does not reduce LER. Figure 2.10 shows LER for different lithography technologies illustrating that LER variation is almost constant for different technology nodes, which means that for shorter channel lengths, the impact of LER is larger.

The variation in transistor channel length due to LER has a direct impact on several electrical properties of a transistor; however, the most affected parameters are the transistor drive current ($I_D \,\alpha\, 1/L$) and V_{th} [16, 25]. The variation in V_{th} arises due to the exponential dependence of V_{th} on channel length L for short channel devices, mainly due to short-channel effects (SCE) and the drain-induced barrier lowering (DIBL) [16, 25]. Both effects cause V_{th} to change strongly dependent on the channel length L as shown in Fig. 2.11. V_{th} shift due to SCE and DIBL can be modeled as [16, 25]:

$$V_{\text{th}} \approx V_{\text{th0}} - (\zeta + \eta\, V_{\text{DS}})e^{-L/\lambda} \tag{2.2}$$

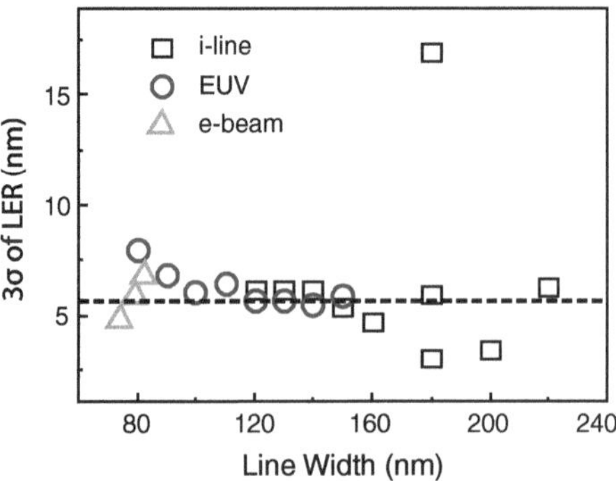

Fig. 2.10 Amplitude of line edge roughness (LER) versus line width for different lithography technologies [33]

where ζ is the SCE coefficient, η is the DIBL effect coefficient, and V_{th0} is the long channel threshold voltage. Therefore, a slight variation in channel length will introduce large variation in V_{th}, as shown in Fig. 2.12.

This type of variation strongly depends on the applied drain-to source-voltage V_{DS}, and the body bias V_{BS}, as shown in Fig. 2.13, because DIBL strongly depends on both V_{DS} and V_{BS} voltages [16, 25]. The roll-off increases as V_{DS} increases. Moreover, as shown in the figure, V_{th} roll-off decreases when forward biasing the body (i.e., V_{BS} positive for NMOS), and vice versa for reverse body biasing (RBB). Therefore, the impact of L variation on V_{th} reduces when applying forward body bias (FBB) [16, 25].

From a circuit modeling approach, the total variation in V_{th} due to RDF, LER, and other static sources of variation, can be formulated as:

$$\sigma^2_{V_{\text{th}}} \approx \sigma^2_{V_{\text{th,RDF}}} + \sigma^2_{V_{\text{th,LER}}} + \sigma^2_{V_{\text{th,other}}} \tag{2.3}$$

2.3.3 Random Telegraph Noise

As transistors continue to become smaller, the impact of single charge perturbation becomes more significant, leading to increased RDF as discussed in earlier sections. In addition to RDF other types of variations arise such as random telegraph noise (RTN), also known as random telegraph signal, RTS [40, 41]. RTN is a random fluctuation in device drain current due to the trapping and detrapping of channel carriers in the dielectric traps at the oxide interface, as shown in Fig. 2.14, which causes variation in V_{th}. The fluctuation in drain current is caused by the change in the number of carriers as well as the changes in surface mobility due to scattering by the trapped charges in the gate dielectric [16]. Both RTN and RDF arise due to discreteness in charges; however RTN significantly differers from RDF in that it is time-dependent, and much fewer charges are involved [40]. Dealing with RTN

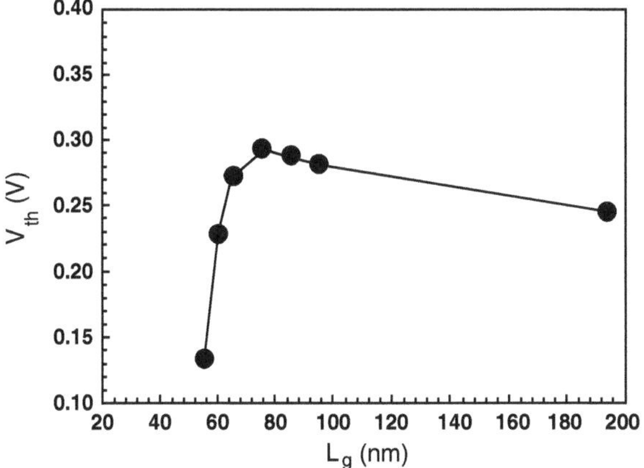

Fig. 2.11 V_{th} versus channel length L in a 90 nm technology [39]. Measured data show the impact of strong short channel effects (SCE) on V_{th} for short channel devices

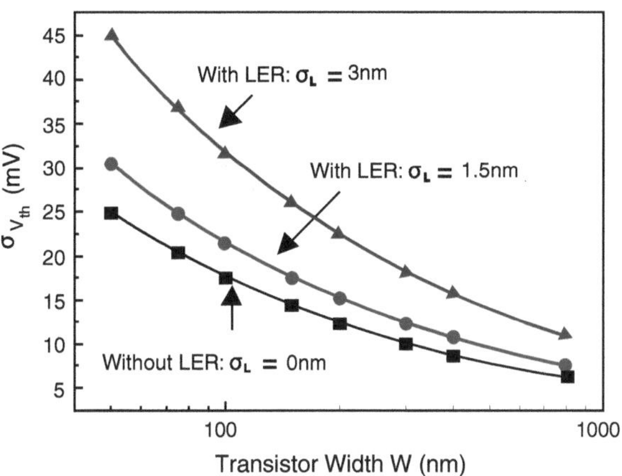

Fig. 2.12 Combined effect of RDF and LER on V_{th} variation. LER increases the variation of V_{th} beyond the values expected from RDF only ($\sigma_L = 0\,\text{nm}$). Results are based on 65 nm predictive models [33]

noise has always been important for analog and radio frequency circuits [34]. With technology scaling, RTN increases due to reduction in the number of channel carriers caused. RTN may become a serious issue for SRAM in the near future [40, 41].

The impact of RTN on V_{th} variations can estimated as follows:

$$\Delta V_{\text{th,RTN}} = \frac{q}{W_{\text{eff}} L_{\text{eff}} C_{\text{ox}}} \tag{2.4}$$

where q is the elementary charger, L_{eff} and W_{eff} are the effective channel length and width, respectively, and C_{ox} is the gate capacitance per unit area. Equation (2.4) shows that $\Delta V_{\text{th,RTN}}$ is inversely proportional to the device area, and can therefore become a serious concern for highly scaled technologies. As shown earlier, $\Delta V_{\text{th,RDF}}$ variation due to RDF is inversely proportional to the square root of device area, while

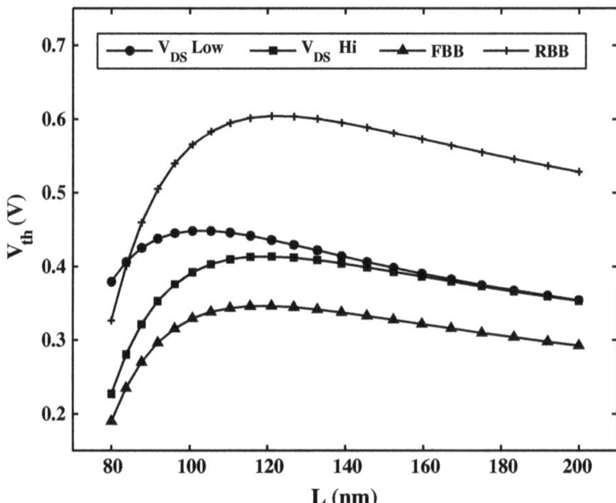

Fig. 2.13 Simulated V_{th} versus channel length L showing V_{th} roll-off under low and high V_{DS} and forward (FBB) and reverse body bias (RBB)

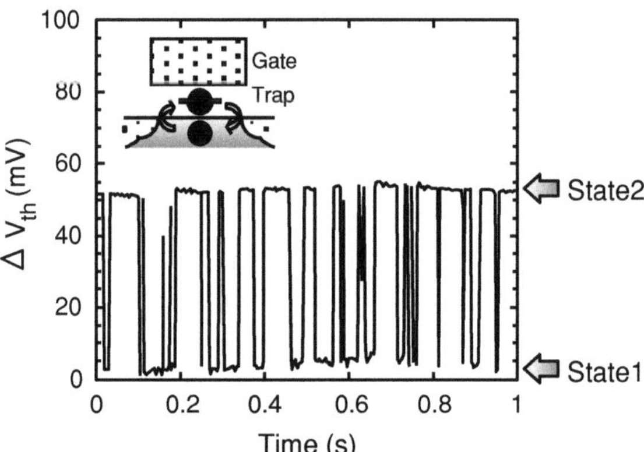

Fig. 2.14 Random telegraph noise (RTN) V_{th} variation is caused by trapping and detrapping of charges in the channel [41]

$\Delta V_{\text{th,RTN}}$ is inversely proportional to device area. Therefore, with technology scaling, variation due to RTN is expected to exceed the RDF component [29, 41, 42].

Detailed measurements in 22 nm generation have shown that $\Delta V_{\text{th,RTN}}$ exceeds 70 mV for smaller devices [41]. One of the critical concerns with RTN is that V_{th} variation due to RTN has a non-Gaussian distribution with a long tail, as shown in Fig. 2.15. RTN V_{th} variations are projected to exceed RDF V_{th} beyond the 3σ point; hence, RTN may exceed RDF in design impact [41].

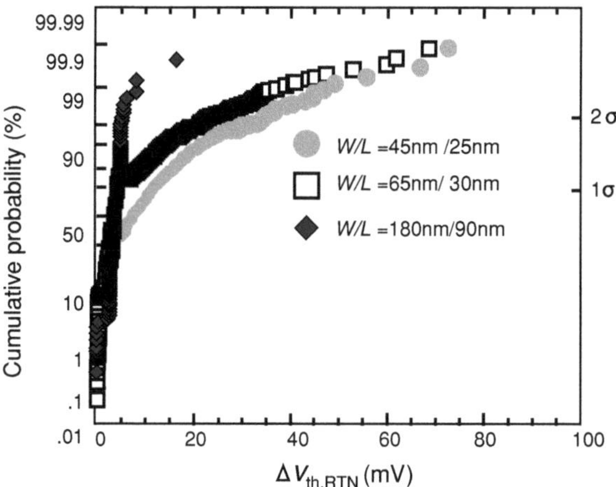

Fig. 2.15 Distribution of V_{th} fluctuation due to random telegraph noise (RTN) in a 22 nm technology, which shows a long-tailed distribution [41]

2.3.4 Time-Dependent Degradation and Aging

In addition to static (time-independent) sources of variations and intrinsic device noise such as RTN, highly scaled devices are degraded over time due to long-term stress and device aging. One of the most critical sources of device aging is negative-bias temperature instability (NBTI), which affects PMOS devices. NBTI causes the magnitude of V_{th} to shift due to the generation of interface traps at the oxide interface when a negative voltage is applied to the PMOS gate for a long period of time. NBTI threshold voltage shift $\Delta V_{th,NBTI}$ due to static stress after time t can be modeled as [43]:

$$\Delta V_{th,NBTI} = A((1+\delta)t_{ox} + \sqrt{C(t-t_0)})^{2n} \tag{2.5}$$

$$A = (\frac{qt_{ox}}{\varepsilon_{ox}}) \sqrt[\frac{1}{2n}]{K^2 C_{ox}(V_{gs}-V_{th})\exp(\frac{V_{gs}-V_{th}}{t_{ox}E_0})} \tag{2.6}$$

$$C = \frac{\exp(-E_a/kT)}{T0} \tag{2.7}$$

where n is the time exponent (1/6 or 1/4 depending on the NBTI model), q is the electron charge, T is the temperature, k is the Boltzmann constant, and C_{ox} is the oxide capacitance per unit area. A is proportional to the hole density and depends exponentially on temperature, and T_0, E_a and E_0, K are constants that are extracted by fitting the measured data. Equation (2.7) shows that $\Delta V_{th,NBTI}$ shift strongly depends on temperature (via C) and oxide thickness t_{ox} (via A) [43]. Due to the reduction of t_{ox} with technology scaling, the impact of NBTI on V_{th} increases.

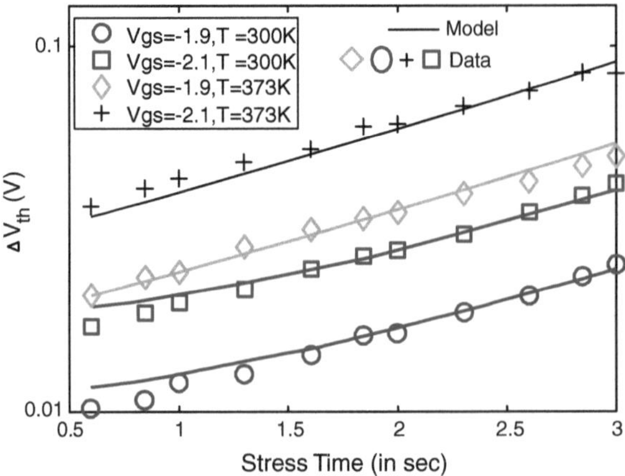

Fig. 2.16 Measured V_{th} degradation under static NBTI for different temperatures and V_{gs} for 90 nm technology [43]

V_{th} shift due to NBTI can be simplified as:

$$\Delta V_{th,NBTI} \approx At^n \tag{2.8}$$

where n is the time exponent which has been shown to vary widely (0.14–0.5) [44, 45].

Figure 2.16 shows V_{th} degradation under static NBTI for 90 nm technology at different temperature and voltage conditions. NBTI shift recovers slightly after the stress condition is removed. Models for V_{th} shift that account for recovery and dynamic stress have been developed. It is important to note that recovery time is in the mS to S range, which complicates the process of accurately measuring the NBTI stress effects [46].

For newer technologies using high-K dielectrics, NMOS devices suffer from a similar reliability concern due to positive bias temperature instability (PBTI) [47]. In addition, there are other sources of time-dependent device degradation due to aging such as the hot carrier effect (HCI) [16, 25] and time-dependent dielectric breakdown (TDDB).

2.3.5 Other Sources

While random dopant fluctuation and LER are currently the dominant sources of device variations, there are several other sources which may become significant in the future technologies. Below we list other sources of device variations:

- **Oxide Charges Variation**: Interface charges can also cause V_{th} variation although their effect is not significant in nitrided gate oxides [24]. The recent adoption of high-K gates to reduce gate-tunneling leakage current may worsen oxide charge variations [24]. In addition, oxide charge variations can introduce mobility fluctuations, as they increase scattering in a transistor channel.

- **Mobility Fluctuation**: Variations in a transistor's drive current can also be caused by mobility fluctuation. Mobility fluctuation can arise from several complex physical mechanisms such as fluctuations in effective fields, fixed oxide charges, doping, inversion layer, and surface roughness [24].Throughout their shared dependence on many physical variation mechanisms, mobility variation shows a certain level of correlation with V_{th} variations. Device measurements show this correlation to be small [48]. Therefore, mobility variations and V_{th} variations are typically assumed to be independent in circuit modeling [48].
- **Gate Oxide Thickness Variation**: Any variation in oxide thickness affects many electrical parameters of the device, especially V_{th}. However, oxide thickness is one of the most well-controlled parameters in MOSFET processing. Therefore, it does not affect V_{th} variation significantly.
- **Channel Width Variation**: Due to lithography limitations, transistor channel width also varies due to lithography (similar to LER). Width variations can cause V_{th} variations in devices that suffer from narrow-width effects (NWE) [16]. However, since W is typically 2–4 times larger than L, the impact of W variation on V_{th} is considered to be smaller than the impact due to L variation [16].

2.3.6 Highly Scaled Devices: FinFET

At the time of writing this book, FinFET or trigate transistors started showing up in production [49, 50]. FinFET is considered a fundamental change in CMOS technology because the device moved from being planar to becoming a 3D structure as shown in Fig. 2.17. FinFET technology reduces SCE and significantly reduces leakage current, which allows lower operating voltage [49, 50]. Due to the improved electrostatic integrity in FinFET, the channel control via the gate can be maintained with lower channel doping. As the fin thickness is reduced, the FinFET becomes fully depleted, the channel can be undoped which provides large reduction in V_{th} fluctuation due to RDF. Figure 2.18 shows a comparison between V_{th} variation for 45 nm bulk and FinFET. With a 6X reduced chancel doping, FinFET can reduce V_{th} variations by more than 60 %.

While FinFET structure helps reduce the RDF contribution to variation, it adds new sources of variation such as fin thickness variations, as shown in Fig. 2.19. Moreover, the FinFET width is determined using fin height, which is constant for a given technology, so the transistor width is quantized and cannot be adjusted in fine steps as in bulk technologies. Width quantization is a major concern for SRAM bitcell design since the current ratios of pass-gate (PG), pull-down and pull-up can only take integer values as shown in Fig. 2.17, preventing optimization of read and write stability through sizing as in conventional bulk technologies. Therefore, read and write assist techniques are necessary for FinFET SRAM [49, 50].

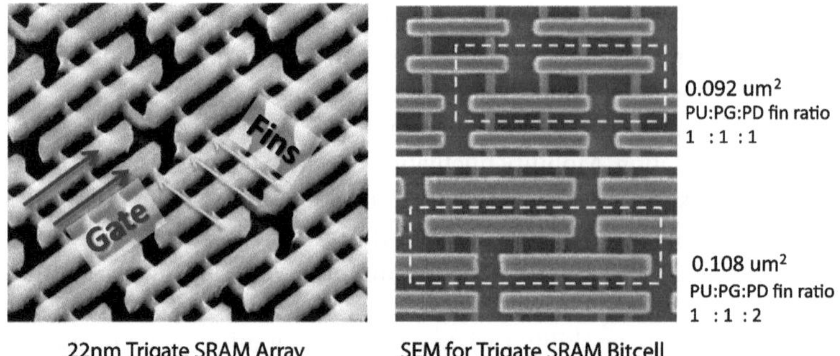

Fig. 2.17 22 nm FinFET SRAM array and the layout of two different FinFET bitcells. Width quantization limits the ratios of pull-up, pass-gate, and pull-down devices to integer values [49, 50]

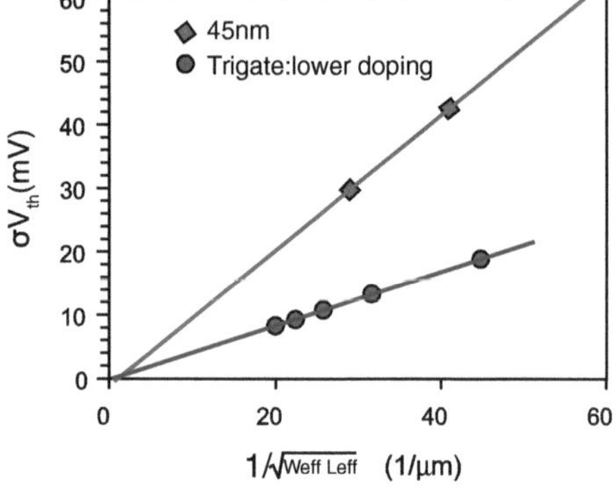

Fig. 2.18 Comparison between V_{th} variation due to RDF for conventional bulk technology (45 nm) and FinFET (trigate) using a lightly doped channel. V_{th} variation reduces significantly for FinFET [51]

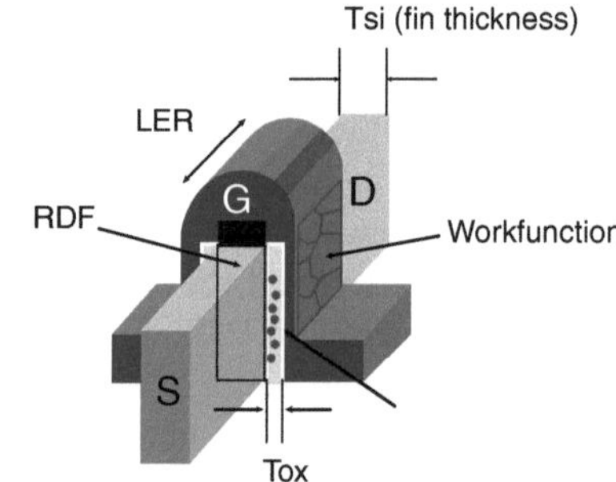

Fig. 2.19 Sources of V_{th} variation in highly scaled FinFET including RDF, LER, fin thickness (T_{si}), oxide thickness (T_{ox}), and workfunction variations [52]

2.4 Interconnect Variability

Similar to the sources of variations that alter device characteristics, several factors affect the interconnects. The mains sources of variations in interconnects include [14]:

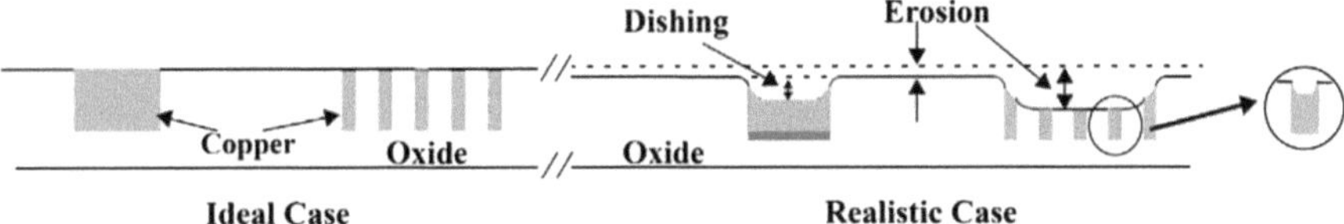

Fig. 2.20 A schematic cross-section of interconnect showing dishing and erosion impact on metal height

1. **Line Width and Line Space**: Deviations in the width of patterned lines arise primarily due to photolithography and etch dependencies. At the smallest dimensions, which typically occur at lower metal levels, proximity and lithographic effects are likely most important. However, at higher metal levels, aspect ratio-dependent etching, which depends on line width and local layout, are generally more significant. Variations in line width directly impact line resistance as well as line capacitance [14, 15].
2. **Metal and Dielectric Thicknesses**: In a conventional metal interconnect, the thickness of metal films is usually well controlled, but can vary from wafer-to-wafer and across the wafer. However, in advanced damascene copper interconnect processes, this is not the case. Unlike older aluminum interconnect processes where the metal is patterned and the oxide is polished, the oxide is patterned, and the metal is polished in a damascene process for copper interconnects. Chemical mechanical polishing (CMP) is then used to flatten the topography on the wafer. Because copper and adjacent dielectric are removed from the wafer at different rates during CMP (depending on the density of the surrounding pattern), this creates surface anomalies such as dishing and erosion. Dishing occurs when the copper recedes below the level of adjacent dielectric and erosion is a localized thinning of the dielectric, which normally happens when CMP is applied to an array of dense lines, as shown in Fig. 2.20. The oxide between wires in a dense array tends to be over-polished compared to the nearby areas of wider insulators. Dishing and oxide erosion are layout dependent; they are problematic in wide lines and dense arrays, respectively. In damascene processes with copper interconnects, dishing and erosion can significantly impact the final thickness of patterned lines, with line thickness losses of 10–20 % leading to higher resistance and capacitance variations [14, 15].
3. **Contact and Via Size**: Contact and via sizes can be affected by variations in the etching process, as well as layer thickness variations. Depending on the via or contact location, the etch depth may be substantially different, resulting in different sizes of the lateral opening. Such size differences can directly change the resistance of the via or the contact [14].

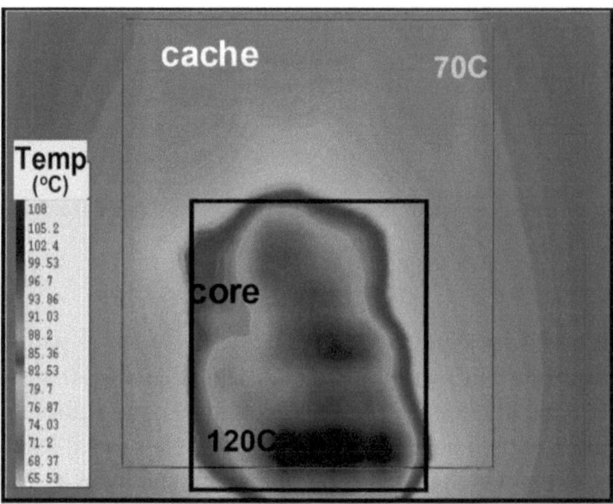

Fig. 2.21 Thermal image showing within die temperature variation for a microprocessor [18]. Hot spots with temperatures as high as 120 °C are shown

2.5 Environmental Variability

In addition to static process variations, environmental factors, which are typically dynamic, also cause variation in the circuit operation. These include variations in power supply and temperature of the chip or across the chip [13, 17, 19].

Variation in switching activity across the die results in uneven power dissipation across the die and uneven supply voltage noise (droops or bumps). A reduced power supply lowers drive strengths, degrades speed for digital logic, and increases SRAM failures [18, 53].

WID temperature fluctuations have always been a major performance and packaging challenge, especially for high-performance processors. Temperature fluctuations are problematic because both device and interconnect have temperature dependencies which cause performance to degrade at higher temperatures. Moreover, temperature variation across communicating blocks on the same die can cause performance mismatches, which may lead to functional failures [18]. As shown in Fig. 2.21, the temperature difference between the core and the cache of a microprocessor can be as high as 50 °C.

Leakage currents, especially subthreshold leakage, strongly depend on temperature; leakage power increases at higher temperatures [14, 17, 18]. In the meantime, higher leakage power causes die temperature to rise. This type of positive feedback may cause thermal runaway where leakage currents, and temperature continue to increase until failure [15].

Both supply and temperature variations depend on the work load of the processor and are thus, time-dependent. However, identifying worst-case conditions for temperature and supply is very difficult [15]. Therefore, designers often focus on minimizing temperature and supply variations as much as possible; for example, ensuring that the voltage drop on the power grid is always less than 10 % of the nominal supply voltage, and by adding large decoupling capacitors [15, 17].

2.6 SRAM Failure Mechanisms

Due to their small size and high density, several factors can cause SRAM bitcells to fail. Figure 2.22 shows various SRAM failures such as stability fails, radiation induced soft errors, and hard fails.

2.6.1 Bitcell Stability Failures

The effect of technology scaling on SRAM failure probability is shown in Fig 2.23 for advanced technology nodes spanning 130 nm down to 45 nm. Traditional hard fails due to defect density decrease due to the reduction of bitcell size and improvement in defect density. However, as the bitcell size is reduced by about 50 % in each technology node, process variations increase significantly and become the dominant cause of bitcell failure [3, 6, 7]. This increase in SRAM failures has a strong impact on the overall product yield due to the high memory densities on chip. Moreover, lower V_{DD} operation becomes limited by the SRAM minimum supply voltage V_{min} due to the sensitivity of stability failures to supply voltage.

There are four main parametric failure mechanisms (also known as SRAM stability failures) [5, 9, 54–56]:

1. read access failure;
2. read stability or read disturb failure;
3. write failure;
4. hold or retention fail.

These failures are parametric in nature since they affect the memory operation under specific conditions. For example, these failures mostly appear as low V_{DD}, while they can be recovered at higher supply voltages. Therefore, these failure mechanisms become the limiting factor for SRAM supply voltage scaling [44, 57, 58].

2.6.1.1 Read Access Failure

During read operation, the wordline (WL) is activated for a small period of time determined by the cell read current, bitline loading (capacitance) as shown in Fig. 2.24. The content of a cell is read by sensing the voltage differential between the bitlines using a sense amplifier. For successful read operation, the bitlines precharged to V_{DD} should discharge to a voltage differential value which can trigger the sense amplifier correctly. Read failure occurs if bitcell read current (I_{read}) decreases below a certain limit, which often results from an increase in V_{th} for the PG or pull-down (PD)

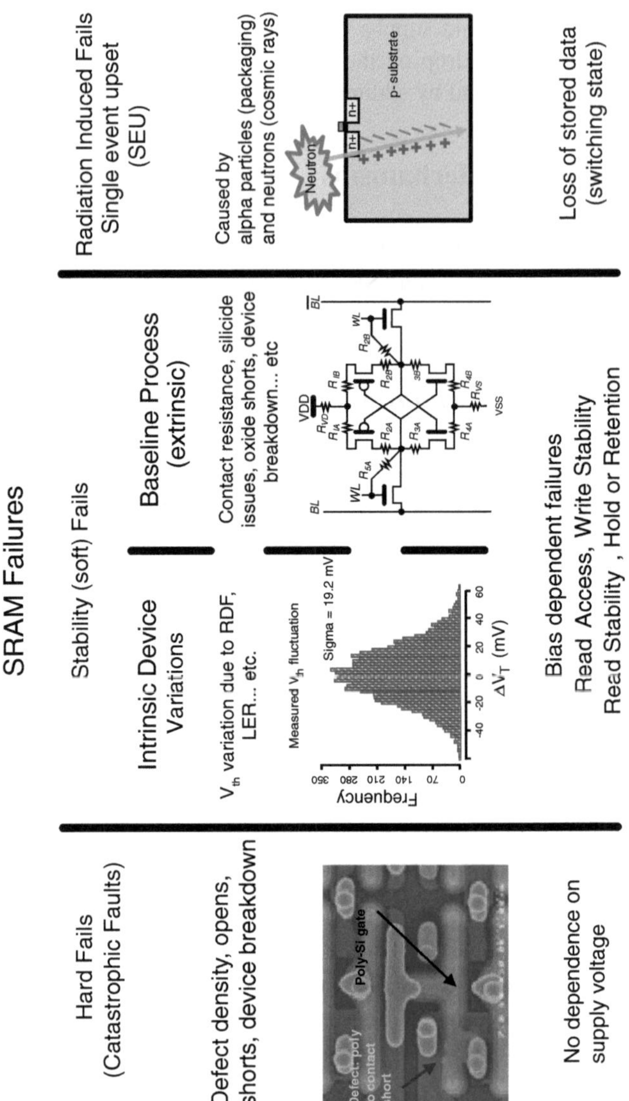

Fig. 2.22 Different types of SRAM failures

transistors, or both. This decrease in I_{read} reduces the bitline differential sensed by the sense amplifier, resulting in incorrect evaluation. This failure can also occur due to large offset affecting the sense amplifier. This type of failure decreases memory speed [5, 9, 54, 59] because the WL activation time is about 30 % of memory access time [60]. Analysis and statistical simulation of read access yield will be discussed in Chap. 5.

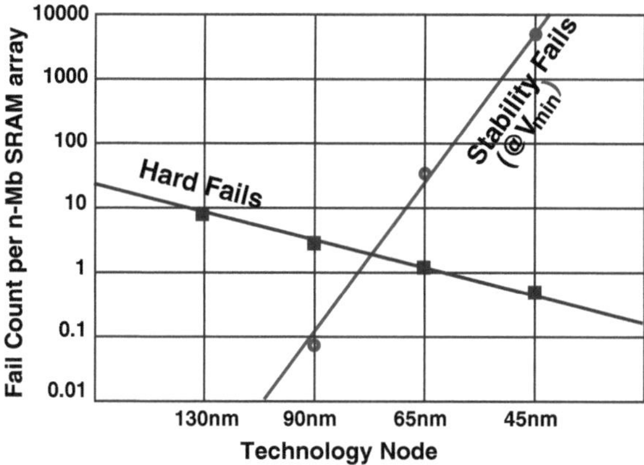

Fig. 2.23 SRAM hard fails and stability fails versus technology scaling [7]

2.6.1.2 Read Stability Failure

SRAM cells are designed to ensure that the contents of the cell are not altered during read access, and the cell can quickly change its state during write operation. These conflicting requirements for read and write operations are satisfied by sizing the bitcell transistors to provide stable read and write operations [5, 9, 54].

In read operation, an SRAM bitcell is most prone to failure. After the WL is enabled, voltage at the internal storage node storing a zero (Q) slightly rises due to the voltage divider between the PG transistor (PG1) and the pull-down (PD1), as shown in Fig. 2.24. If the voltage at Q rises close to the threshold voltage of the adjacent pull-down, PD2, the cell may flip its state. Therefore, stable read operation requires that PD1 should be stronger than PG1. Read stability failure is exacerbated by process variations, which affect all the transistors in the bitcell [5, 9, 54]. To quantify the bitcell's robustness against this type of failure, static noise margin (SNM) is one of the most commonly used metrics [61]. A read stability failure can occur if the bitcell cannot hold the stored data, in which case SNM is zero [5, 54, 61].

Read stability failure can occur any time the WL is enabled even if the bitcell is not accessed for either read or write operations. For example, in half-selected bitcells, the WL is enabled while the bitlines column is not selected (the bitcells are not actively accessed for read or write). These bitcells experience a dummy read operation because the bitlines are initially precharged to V_{DD}, and the bitlines are discharged after the WL is enabled, hence, the bitcells become prone for read stability failure. Dealing with read stability failures is one of the biggest challenges for SRAM design and has been extensively studied [5, 9, 54]. Circuit techniques to deal with read stability failures will be discussed in Chap. 3.

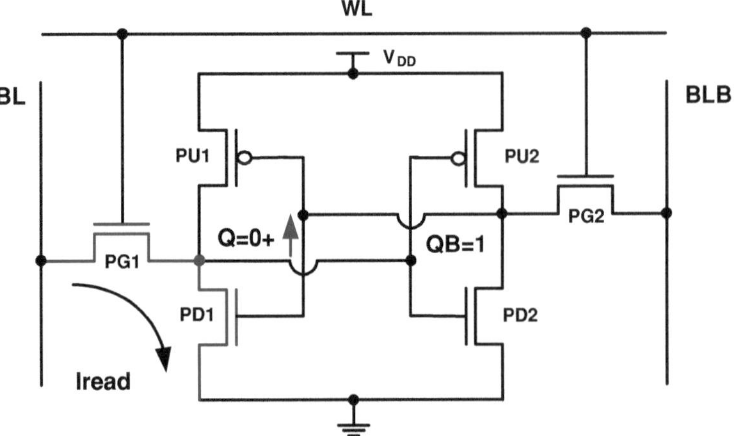

Fig. 2.24 Bitcell in read operation

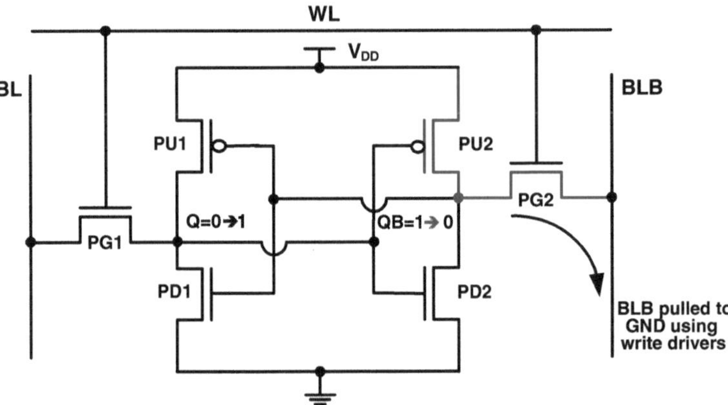

Fig. 2.25 Bitcell in write operation

2.6.1.3 Write Stability Failure

The ability of the bitcell to be written correctly is referred to as write stability or write margin. In write operation, BLB is pulled to zero by the write driver, while WL is enabled, as shown in Fig. 2.25. Therefore, the NMOS PG2 is turned ON, which results in a voltage drop in the storage node QB holding data 1 until it falls below $V_{DD} - V_{th}$ for the PU1, where the positive feedback action begins. For stable write operation, PG2 should be stronger than PU2. Due to WID variations, the pass gate cannot overcome the pull-up transistor, resulting in a write failure [54, 62]. Write failure can also happen if the WL pulse is not long enough for the bitcell to flip the internal nodes (dynamic failure).

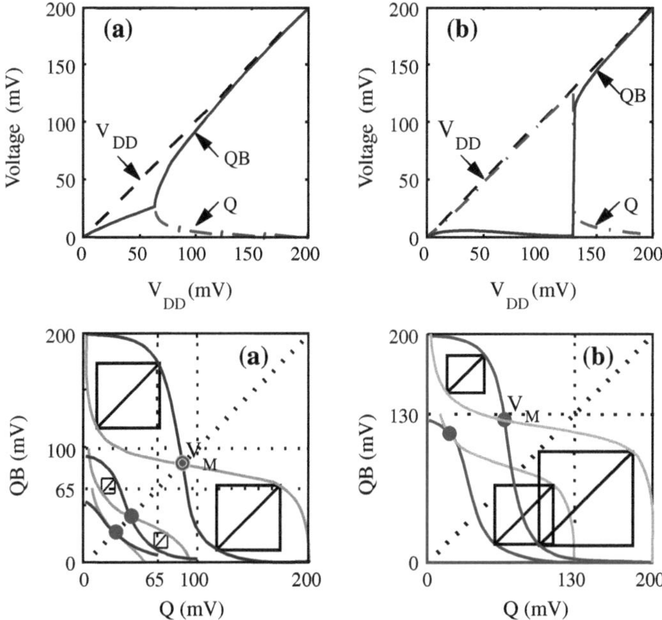

Fig. 2.26 Data retention failure mechanism. *Upper figures* show the bitcell internal node voltages Q and QB for **a** balanced and **b** imbalanced cell as V_{DD} is reduced. *Lower figures* show the voltage transfer characteristics (VTC) of **a** balanced and **b** imbalanced cell with varying V_{DD}. V_M is the trip point of the VTCs. The imbalanced bitcell has higher data retention voltage (DRV) than the balanced case because of asymmetry in the VTCs caused by variations [58]

2.6.1.4 Data Retention Failure

Reducing supply voltage (V_{DD}) is one of the most effective techniques to reduce both static and dynamic power consumption for digital circuits [63]. In SRAM, the data retention voltage (DRV) defines the minimum V_{DD} under which the data in a memory is still preserved. When V_{DD} is reduced to DRV, all six transistors in the SRAM cell operate in subthreshold region, and therefore are strongly sensitive to variations [57, 58].

DRV depends strongly on WID variations in the bitcell inverters, which may cause the bitcell to be imbalanced. This imbalance can be examined using SNM in standby (WL is disabled) as shown in Fig. 2.26. If the bitcell is asymmetric due to WID variations, the bitcell tends to have a higher DRV than in the symmetric case. This can be explained using SNM, where DRV voltage can be defined as the voltage when hold SNM is equal to zero. In the symmetric case, both SNM high (upper left square) and SNM low (lower right square) decrease symmetrically to zero. However, in the case of asymmetric bitcell shown in Fig. 2.26, SNM low is always larger than SNM high, and the bitcell DRV is limited by the SNM high case. Therefore, variations increase the bitcell DRV because they increase the asymmetry [57, 58].

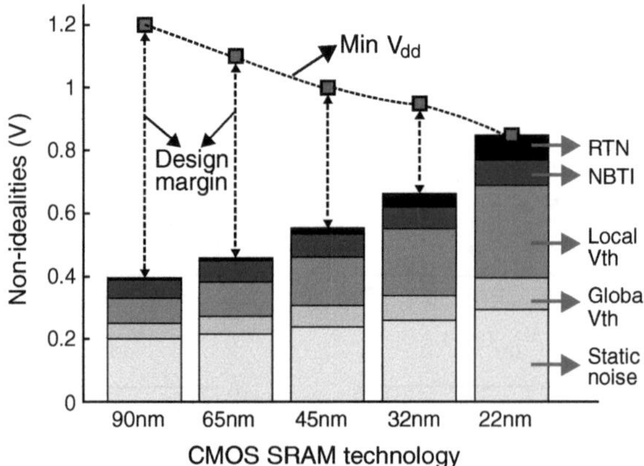

Fig. 2.27 Reduction of SRAM design margin with technology scaling due to increase in variations [64]

2.6.2 Impact of Variations on SRAM V_{min}

Variations affect SRAM operations and set the limit of the SRAM minimum supply voltage $V_{\min}$. Due to the increase in variations with technology scaling, the operation margin of SRAM decreases, as shown in Fig. 2.27, making the SRAM $V_{\min}$ the limiting factor of voltage scaling in newer technologies. Measured distribution of SRAM $V_{\min}$ is shown in Fig. 2.28. The distribution shows the large spread of $V_{\min}$ due to the impact of RDF and LER on the bitcell stability. Figure 2.29 shows the impact of WID variations on SNM, which translates into larger $V_{\min}$ spread.

As discussed earlier, NBTI and PBTI cause V_{th} of PMOS and NMOS devices to increase, which decreases the bitcell stability [66–68]. Figure 2.30 shows that as SNM decreases due to NBTI, the probability of read disturb increases. The increase in read disturb failures causes an increase in the $V_{\min}$ since higher voltage is required to recover the degradation in SNM. Figure 2.31 shows the shift in SRAM minimum supply voltage $V_{\min}$ due to burn-in stress. As the burn-time increases, $V_{\min}$ shifts towards higher voltage, which can be attributed to the increase in PMOS V_{th} due to NBTI and the associated reduction in SNM. For some parts, stress causes $V_{\min}$ to decrease, as shown in the negative values of $V_{\min}$ shift, which may be explained by improvements in bitcell write margin since PMOS pull-up becomes weaker. However, even for read-limited memories, $V_{\min}$ shift may be negative; and there is little correlation between initial $V_{\min}$ and its final value after stress, which is due to the combined impact of static variations (RDF, LER) and NBTI degradation [44].

Depending on the bitcell size and stability optimization, NBTI's impact on $V_{\min}$ can vary. Figure 2.33 shows that for read-limited bitcells, NBTI causes the $V_{\min}$ distribution to move towards higher voltage. However, for write-limited bitcells, the $V_{\min}$ distribution to lower voltage, since the strength of the pull-up device decreases due to NBTI, which improves write margin [44, 67, 69].

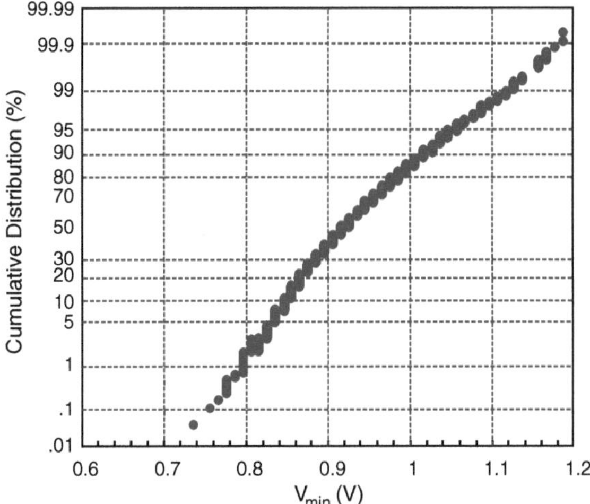

Fig. 2.28 Measured V_{min} distribution for a 2 Mbit SRAM in 65 nm technology. V_{min} varies widely due to random variations, causing read and write stability failures [65]

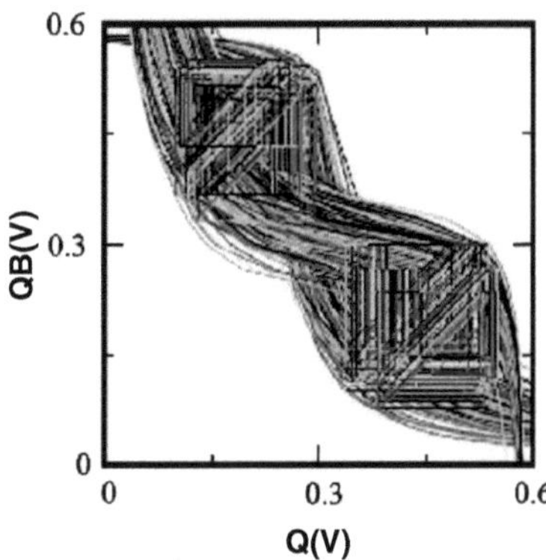

Fig. 2.29 Measured SNM butterfly curves for 512 bitcells in a 65 nm technology node showing the strong impact of WID variations on SNM [6]. A large spread in butterfly curves causes SNM to be unsymmetrical and increases the probability of bitcell failure

Another important aspect of NBTI degradation is time dependence and its impact on V_{min}. As shown in Fig. 2.32, increasing burn-in increases NBTI shift, which increases V_{min}. The shift in PMOS pull-up V_{th} due to NBTI causes the SRAM V_{min} to increase linearly [47]. V_{min} increases by 110 mv after the first 8 h of stress, and increases by an additional 30 mV after 160 h. The initial large shift and eventual saturation of V_{min} reflects the time dependence of $\Delta V_{th,NBTI}$ which follows a fractional power law $\Delta V_{th,NBTI} \propto t^n$, where $n < 0.5$ as explained earlier [44, 45, 67]. Simulation studies for high-K dielectrics show that SRAMs are more susceptible to V_{min} stability problems due to the combined effects of PMOS NBTI and NMOS PBTI. Guard-banding or screening for time-dependent shift in SRAMs is a challenge.

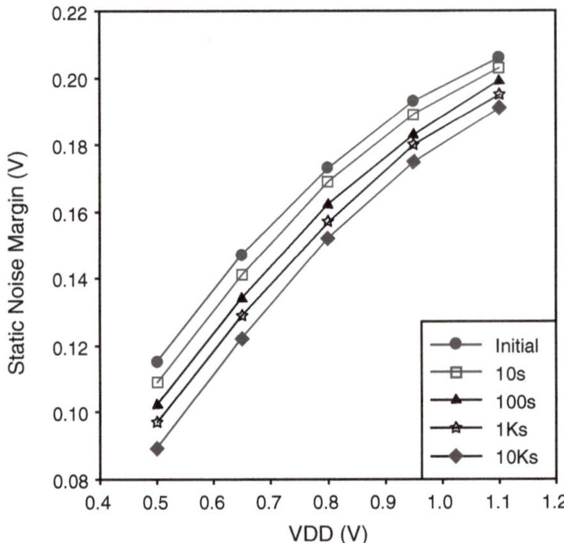

Fig. 2.30 Measured SNM versus V_{DD} for different stress times in a 65 nm node. SNM monotonically decreases with stress time due to NBTI degradation [67]

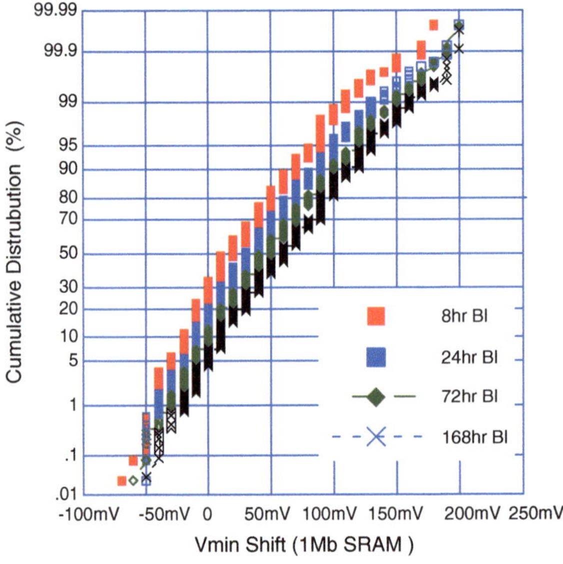

Fig. 2.31 Measured V_{min} shift distribution of a 1 Mb SRAM for different stress/burn-in durations in a 65 nm technology [70]

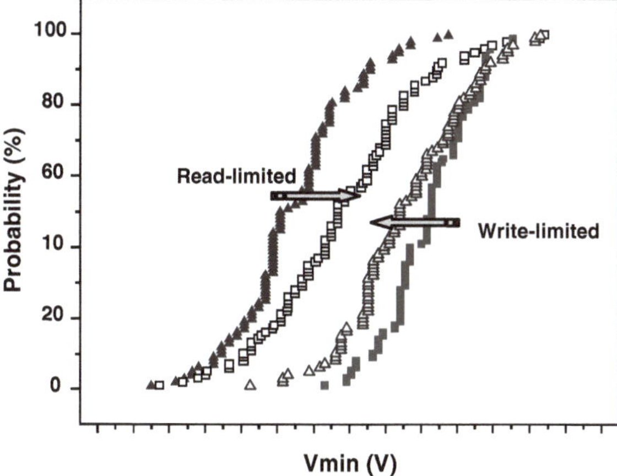

Fig. 2.32 Minimum SRAM supply voltage (V_{min}) distribution for read-limited and write-limited bitcells before (*solid markers*) and after NBTI stress (*hollow markers*) [44]

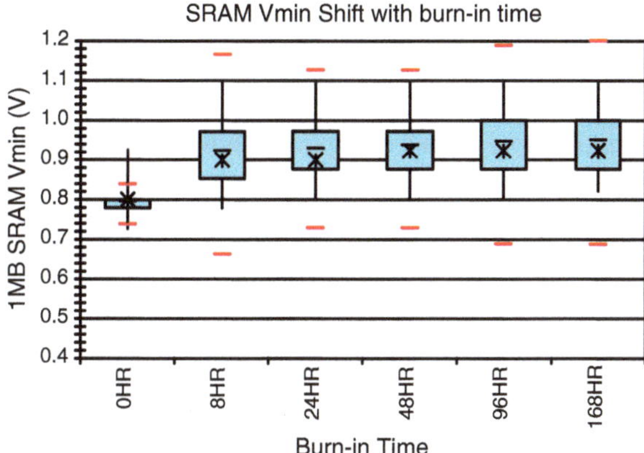

Fig. 2.33 Measured V_{min} shift versus burn-in time in a 65 nm technology [70]

The impact of RTN V_{th} variation on SRAM operation and V_{min} has been researched extensively in the last few years [40, 41, 71–76]. As shown in Fig. 2.34, in smaller bitcells, RTN causes a greater increase in V_{min} shift due to larger $\Delta V_{th,RTN}$, as expected from Eq. (2.4). RTN can increase V_{min} by 50–100 mV for an error-free lifetime condition [74, 75] which is a significant portion of the SRAM V_{min} budget. However, other research predicts that RTN V_{min} degradation becomes less significant in larger arrays [75]. Due to the time dependence of RTN, it is difficult to screen for RTN failures by ordinary functional tests. Design margining techniques as well as accelerated tests have been proposed to deal with RTN issues [40, 76]. Recent work also suggests that there is a relationship between RTN and NBTI, so, margining techniques need to account for both effects simultaneously [77].

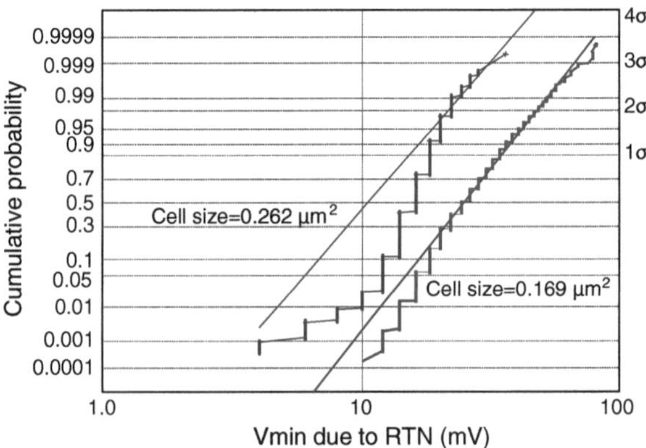

Fig. 2.34 ΔV_{min} distribution due to RTN for two bitcell sizes, which shows that smaller bitcell has larger V_{min} shift since RTN is inversely proportional to FET size [74]

2.6.3 Radiation-Induced Soft Errors

SRAMs are susceptible to dynamic disruptions known as single event upsets (SEU) [14]. SEUs arise due to energetic radiation (Alpha particles or cosmic rays) that hits the silicon substrate and generates free electron–hole pairs, which can affect the potential of bitcell storage nodes and flip the stored data. To determine the susceptibility of SRAM to SEUs, the critical charge that can cause a storage node to be disrupted (Q_{crit}) is calculated. However, with technology scaling, SRAM junction capacitance, cell area, and supply voltage are all scaled down. These reductions have opposing effects on Q_{crit} and the collected charges. However, the combined effect causes the rate of SEUs, or SER, to saturate or slightly decrease with technology scaling [78–80], as shown in Fig. 2.35. The reduction in single-bit SER does not necessarily translate into a reduction in the overall system failure rate due to the rapid growth in embedded SRAM density. In fact, SRAM systems failure rates are increasing with scaling and have now become a major reliability concern for many applications [3, 78, 79]. Moreover, process variations lead to large variation in Q_{crit} which also affects SER [81]. Other research, shows that the aging due to NBTI, oxide breakdown, and hot carriers has negligible impact on SER [82].

To mitigate soft errors, several radiation-hardening techniques can be implemented through process technology (e.g., SOI technology), circuit design (e.g., adding feedback capacitors, larger transistors, columns/words interleaving) and architecture (e.g., parity, error correction codes) or a combination of all these techniques [78]. The SER budget for chips or systems is typically set based on target market requirements. For example, for single user, single chip applications as in mobile phones, it is acceptable to have an average failure rate of about one error every two years due to SER. On the other end of applications spectrum, the same

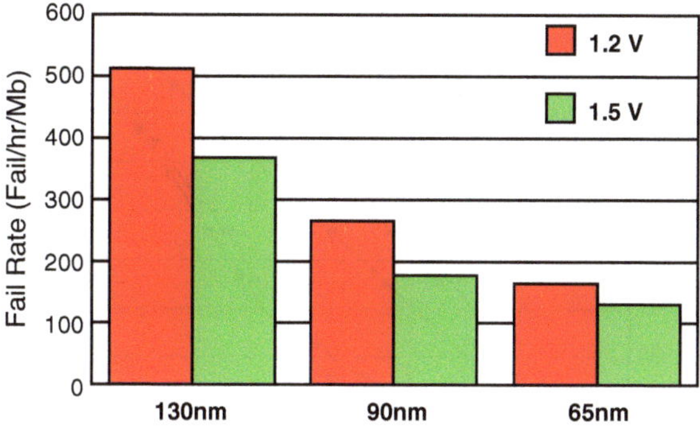

Fig. 2.35 SER failure rate for different technology nodes. SER per Mbit decreases with scaling [3]

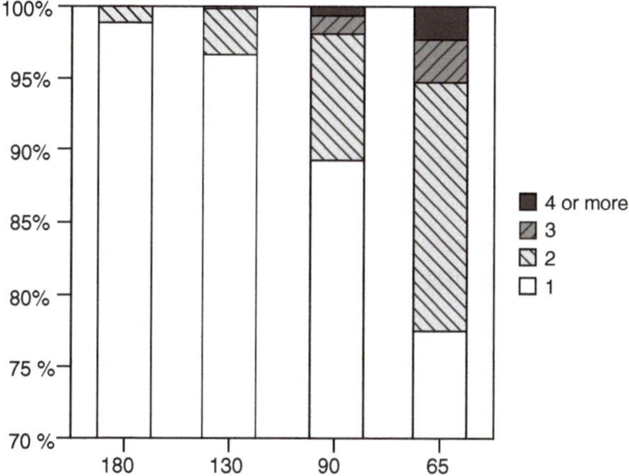

Fig. 2.36 Multicell upsets (MCUs) for different technology nodes, which shows the increase in MCUs with scaling [85]

failure rate is not acceptable for high reliability systems utilizing hundreds of chips as in telecom base stations or servers [78, 79, 83].

In addition to SEUs, multi-cell upsets (MCUs) are becoming a concern for high integrity systems. An MCU consists of simultaneous errors in more than one memory cell and is induced by a single event upset [3, 84, 85]. Risk of MCUs is typically minimized by column interleaving and ECCs. Figure 2.36 shows the increase in MCUs with technology scaling , which results from closer physical proximity of neighboring cells. The increase in MCUs may become a critical reliability concern with future technology scaling [3, 84, 85].

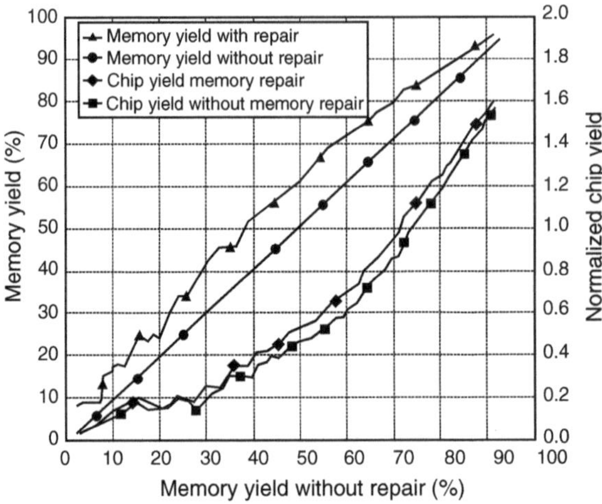

Fig. 2.37 Impact of memory repair on memory and chip yield [88]

2.6.4 Hard (Catastrophic) Fails

Hard fails due to physical defects, also called catastrophic fails, can cause permanent damage for memory and digital circuits. Physical defects include a wide range of possible defects such as voids, shorts, metal bridges, missing contacts or vias, oxide pin holes and many others [14]. Because memories are designed with aggressive design rules, memories tend to be more sensitive to manufacturing defects than other logic circuits in the chip [1]. Figure 2.23 shows that hard fails decrease with process technology due to lower device area, while soft fails due to intrinsic variation increase [3].

SRAM employs redundancy in rows, columns or banks, allowing replacement of defective elements, which improves yield significantly [1, 14, 86]. Historically, this type of repair capability was implemented to address hard fails. However, nowadays, memory redundancy is also used to recover from yield loss due to bitcell stability failures [54, 87].

Figure 2.37 shows how memory repair can improve both memory and chip yield significantly [88, 89]. Memory repair can be used to enhance yield by 5 to 20 %, depending on the type of redundancy used, which translates into large improvement in the chip yield. Several yield models have been proposed to analyze the impact of hard defects on memory and chip yield [90, 91]. In addition, yield simulation using inductive fault analysis (IFA) and critical area analysis (CAA) are used to estimate memory yield [90, 91].

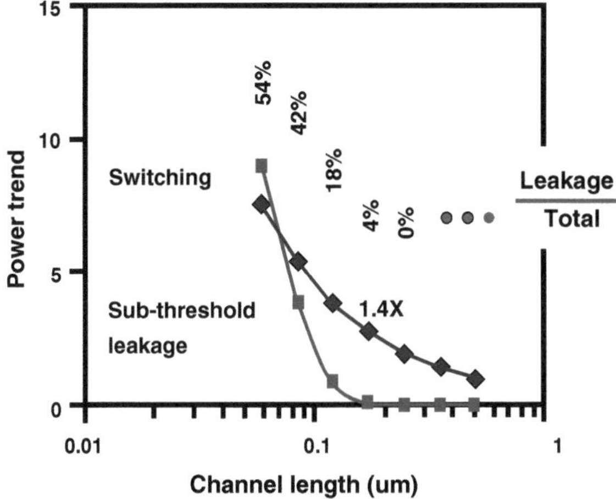

Fig. 2.38 Dynamic and static power versus technology scaling, showing the exponential increase in leakage power in smaller technologies [94]

2.7 Techniques to Deal with Variability for Logic Circuits

In this section, we review state-of-the-art research studying the increase in variability for logic circuits. While the focus of this book is about SRAM, it is instructive to look at the impact of variations on logic circuits to highlight the similarities in the variation-tolerant design approaches.

Performance and power consumption are the most critical metrics for logic circuits. In nanometer devices there are several sources of leakage current, such as subthreshold, gate oxide tunneling, junction band-to-band tunneling (BTBT), and gate-induced drain leakage (GIDL), all of which increase with technology scaling [32, 92, 93]. Therefore, for designs in sub-90nm, leakage is considered a significant part of the total power, and it increases with technology scaling as shown in Fig. 2.38.

The large variability in advanced CMOS technologies increasingly contributes to the total leakage of a chip because leakage depends strongly on process variations [95, 96]. For example, variation in V_{th} introduces a large spread in subthreshold leakage due to the exponential dependence on V_{th}. Similarly, gate-tunneling leakage current is sensitive to oxide variation. The sensitivity of leakage to variations has accentuated the need to account for statistical leakage variations during design [95–97].

For a whole chip, process variations can cause large variations in leakage power. Figure 2.39 shows measured variation of maximum frequency and leakage power for a chip in 130 nm technology [20]. The figure illustrates that the magnitude of leakage variation is much larger than frequency variation (5X leakage spread for a 30 % variation in chip frequency). The chips running at the highest frequency

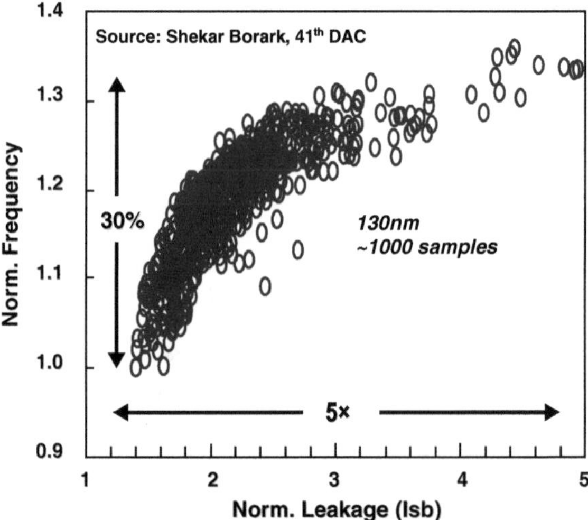

Fig. 2.39 Leakage and frequency variations for a processor in 0.13 μm technology [20]

have a wide distribution of leakage. This excessively large spread in leakage current makes it very difficult to achieve the required speed while meeting power constraints, especially as leakage power increases exponentially with scaling.

Among the chips that meet the required operating frequency, a large fraction dissipate a large amount of leakage power, which makes them unsuitable for usage, degrading the chip yield [18]. The leaky chips have higher frequency because of the inverse correlation between leakage current and circuit delay. For devices with smaller channel length, V_{th} decreases due to short channel effects, and therefore, the subthreshold leakage current increases exponentially. Simultaneously, circuit delay decreases due to the increase in I_{on}, since the overdrive voltage $V_{DD} - V_{th}$ increased. Hence, these chips have a higher operating frequency, but suffer from large leakage power, making them unusable. For the high frequency chips shown in Fig. 2.39, both the mean and standard deviation of leakage current increases considerably, causing yield to decrease substantially [18]. Therefore, there is a crucial need to account for leakage power and its dependence on process variations when analyzing the impact of variability on design techniques [98]. Moreover, variation-tolerant circuit techniques that can reduce variability, and hence reduce leakage power variation, should be designed to improve the yield in advanced CMOS technologies [17, 18, 20].

2.7.1 Circuits

To mitigate variability in logic circuits, considerable research has been done to measure variability and use feedback to mitigate it. These techniques control variability using supply voltage, body bias, or programmable sizing.

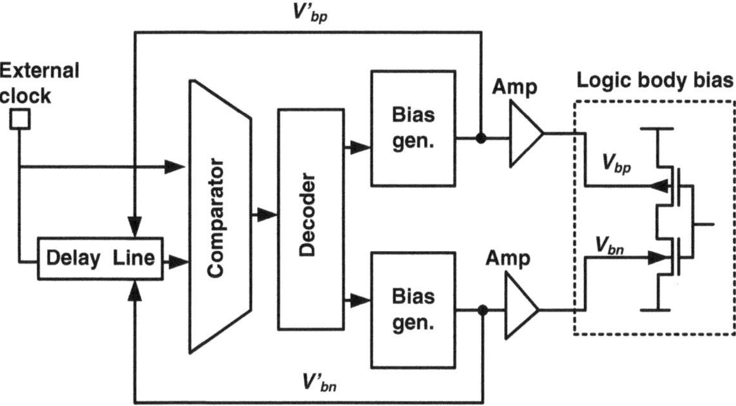

Fig. 2.40 Block diagram of speed-adaptive V_{th} technique [100]

A speed-adaptive body bias technique was utilized in [99, 100] to compensate for variability in microprocessor implemented in 0.2 μm technology. The speed adaptive-V_{th} is composed of a delay line, a delay comparator, a decoder, and body bias generators, as shown in Fig. 2.40. The comparator measures the delay between an external clock signal and an output signal from the delay line and then converts the amount of delay into a register address in the decoder. The generators supply V'_{bp} and V'_{bn} for PMOS and NMOS bodies, respectively, to keep the delay constant by changing the V_{thp} and V_{thn}, respectively. If the speed of the delay line changes due to variation, the comparator output changes, and the generated body bias is modified. The junction leakage and GIDL current determine the maximum reverse-bias voltage, which was set to 1.5 V, while the forward biased was limited to 0.5 V to reduce the subthreshold leakage. In addition, FBB mitigates SCE, and reduces the sensitivity to channel length variation, as was shown earlier in Fig. 2.13.

This technique is efficient in dealing with D2D variations, however, it cannot mitigate WID variations effectively because this technique supplies the same body bias to the entire chip, while WID variations affect different parts of the chip differently.

A similar technique was presented in [101], where again forward and RBB were used to improve performance and decrease leakage, respectively. This adaptive body bias (ABB) allows each die to have the optimal threshold voltage which maximizes the die frequency subject to power constraint. A critical path emulator containing key circuit elements of a process critical path are used to model the effect of body bias on the frequency and leakage of the processor.

This study used multiple delay sensors distributed on the die to get an average body bias that accounts for WID variations [101]. With no body bias, only 50 % of the dies meet the performance and power requirements, mainly in the lowest frequency bin. ABB using only one delay sensor reduced the frequency variation σ/μ from 4 to 1 %, however, a large number of dies still failed to meet leakage constraint.

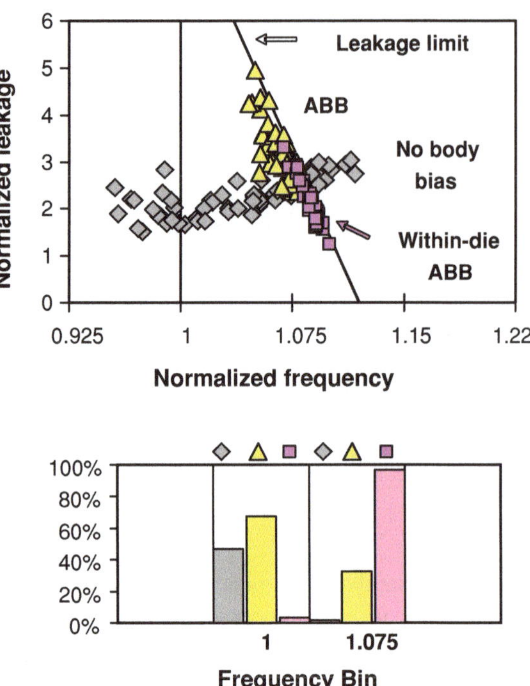

Fig. 2.41 Measured leakage versus frequency scatter for 62 dies in a 150 nm technology (V_{DD} = 1.1V, T=110C), which shows the tighter distributions after utilizing ABB and WID-ABB. In the *lower figure*, the percentage of accepted dies at a certain frequency bin are shown [101]

Using multiple sensors for ABB, the frequency variation reduced to 0.69 % and all dies met the leakage constraint with 32 % in the highest frequency bin.

While the ABB scheme with several sensors considers WID variations in determining the optimum bias combination per die, it still does not completely compensate for these variations, since only a single bias combination is used per die. By allowing different blocks to have their own body bias control (WID–ABB), the effectiveness of the ABB system improves [101]. Figure 2.41 shows the results after using WID–ABB technique. More improvements can be achieved when ABB is combined with adaptive supply voltage V_{DD} [102] However, this improvement comes at the cost of additional area, design complexity, and cost.

Another body biasing technique has been proposed to reduce random fluctuations [103]. In this technique, FBB is applied to the logic circuit blocks, using a body bias generation circuit shown in Fig. 2.42. A current source determines the substrate potential by forward biasing the junction diode. The current source limits the maximum currents that the forward diodes can conduct, and the body potential is self-adjusted by the diode current. Under this self-adjusted FBB (SA-FBB) condition, σV_{th} decreases by 35 %. Interestingly, the improvement achieved using SA-FBB was larger than the improvement using the conventional FBB technique. This improvement may be due to the fact that SA-FBB enables the body bias to become more forward biased compared to conventional FBB because the body voltage is set by

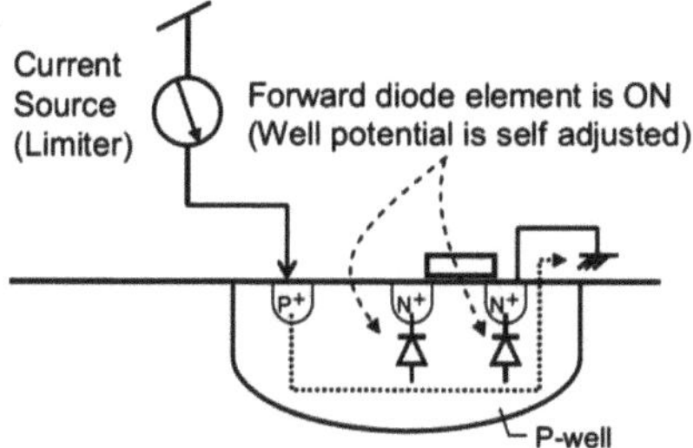

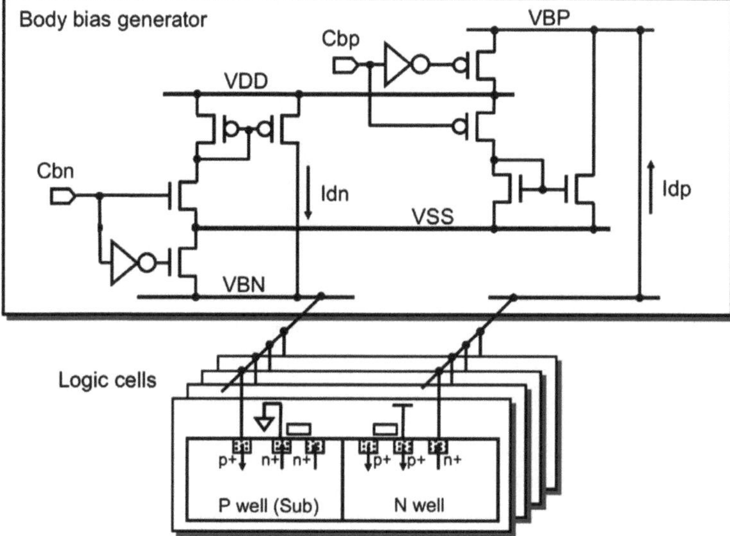

Fig. 2.42 Schematic showing the implementation of self-adjusted forward body bias (SA–FBB) circuit technique [103]

the diode current current used. In contrast, in conventional FBB, the maximum FBB is constrained by stability requirements for the substrate and preventing latch-up.

In addition to voltage and body bias control, programmable sizing has been proposed to improve the robustness of logic circuits. Dynamic circuits are usually used for high-performance gates such as high-speed register files [104]. To prevent the dynamic node from floating and hold it to V_{DD} when none of the pull-downs are evaluated, dynamic circuits use keeper devices [63]. In previous technologies, a small keeper was sufficient to hold the dynamic node to V_{DD}. As technology scales, stronger keepers are required to prevent the dynamic node from collapsing under increasing pull-down leakage levels. In addition, due to the increase in leakage variations, the keeper should be sized up which reduces the the speed of dynamic circuits and limits its advantage over static CMOS [105].

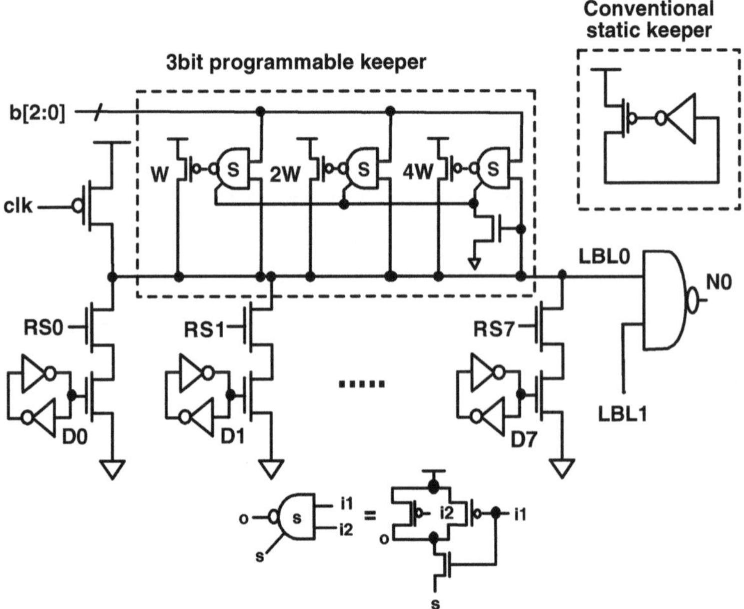

Fig. 2.43 Register file with programmable keeper to compensate for process variations impact on leakage [104]

A variation-tolerant compensation technique for dynamic circuits using programmable sizing is shown in Fig 2.43 [104, 105]. The technique reduces the number of failing dies by 5X compared to conventional designs. An on-chip leakage sensing circuit measures leakage current, and automatically selects the optimal keeper width. In Chap. 3, we will discuss circuit techniques used to mitigate variability in SRAM.

2.7.2 Architecture

Early studies that related variability to architecture were by Bowman et al. [22, 106, 107], which presented a statistical predictive model for the distribution of the maximum operating frequency (FMAX) for a chip in the presence of process variations. The model provides insight into the impact of different components of variations on the distribution of FMAX. The WID delay distribution depends on the total number of independent critical paths for the entire chip N_{cp}. For a larger number of critical paths, the mean value of the maximum critical path delay increases as shown in Fig. 2.44. As the number of critical paths increases, the probability that one of them will be strongly affected by process variations is higher, further increasing the mean of critical path delay. On the other hand, the standard deviation (or delay spread)

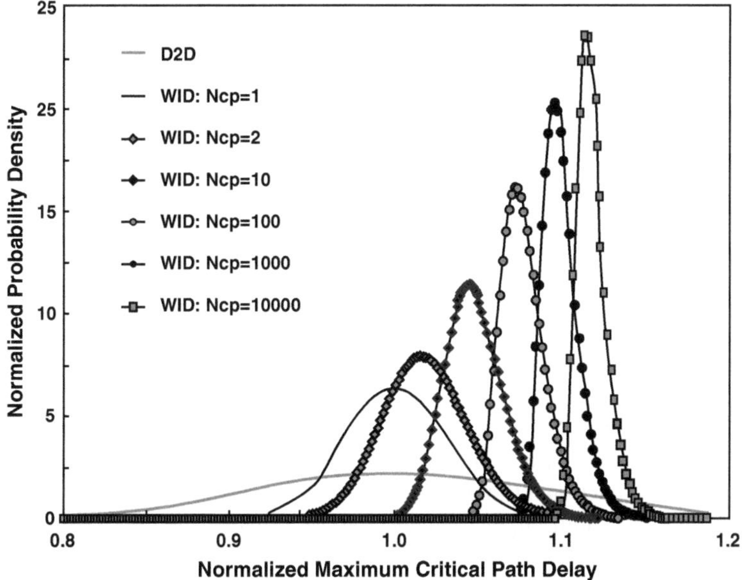

Fig. 2.44 The distribution of maximum critical path delay for different numbers of independent critical paths N_{cp}. As N_{cp} increases, the mean of maximum critical path delay increases [106]

decreases with larger N_{cp}, thus making the delay spread determined mainly by D2D variations. This work revealed that WID variations directly impact the mean of the maximum frequency, while D2D fluctuations impact the variance.

Another factor that affects the delay distribution is the logic depth per critical path. The impact of logic depth on delay distribution depends on whether WID variations are random or systematic. Random WID variations have an averaging effect on the overall critical path distribution (reduces the relative delay variation σ/μ), while systematic WID variations affect all the gates on the path, and thus, increase delay spread [106, 108, 109].

Other variation-tolerant study at the architectural level showed a statistical methodology for pipeline delay analysis [110]. This study emphasized the importance of logic depth in delay variability, and showed that changing logic depth and delay imbalance between stage delays can improve the yield of a pipeline, and showed that the push for high clock speeds using deep pipelining decreases logic depth and increases delay variability [110].

Studies on the impact of variations on low power parallel processing systems show that neglecting WID variation would underestimate the optimum supply voltage that minimizes power consumption for the system [111]. The number of parallel blocks required to meet throughput requirements increases significantly with the increase in WID process variations. As a consequence, the optimum supply voltage that provides the lowest power increases, and therefore, the targeted power parallelism becomes less effective in decreasing power consumptions. An investigation on the impact of parameter variations on multi-core chips showed that WID variation are important for core-to-core granularity, rather than at unit-to-unit granularity [112, 113]. In Chap. 4, we will look at variation-tolerant memory architectures.

2.7.3 Statistical Timing Analysis

Considerable research has been done in the area of CAD tools that are variation-aware. One of the most researched topics in this area is statistical static timing analysis (SSTA) as compared to the well-known static timing analysis (STA) tools [13, 15, 114, 115].

The goal of timing verification is to ensure a chip will operate at a frequency with high yield under the full range of operation conditions. In timing verification, speed (setup time) and functional (hold time) are usually checked to verify that the design will meet the maximum target frequency, as well as provide correct functionality [14].

STA has been used in performance verification for the past two decades. Traditionally, process variations have been addressed in STA using corner-based analysis, where all the gates are assumed to operate at worst, typical, or best-case conditions [13]. This technique is efficient when dealing with D2D (inter-die) variation. However, since WID variation has become a substantial portion of the overall variability, corner-based STA can produce inaccurate predictions that yield low-performance designs, which has motivated the development of SSTA [13].

In SSTA, the circuit delay is considered a random variable and SSTA computes the probability density function (pdf) of the delay at a certain path [13]. The arrival times also become random variables, and therefore, the addition and maximum operations of STA are replaced by convolution and statistical maximum, respectively [13]. Much of the work on SSTA, however, has been in the area of finding efficient algorithms to perform these functions [13]. While SSTA is more appropriate in dealing with WID variations, and can give accurate results without going through the lengthy Monte Carlo simulations, the usefulness of SSTA for timing analysis has been questioned [116]. In Chap. 5, we will look at statistical CAD techniques used in SRAM design.

2.8 Summary

SRAM bitcells continue to scale by 50 % in each technology node, while the density of embedded SRAM increases significantly. At the same time, variability is worsening with technology scaling due to the increase in device variations such as RDF, LER, RTN, and device degradation (NBTI and PBTI). These types of variations have strong impact on the SRAM operation, and increase in the probability of failure in read access, read disturb, write, and hold failures. The increase in variations poses a huge challenge for SRAM design, and determines the lowest operating voltage of the memory $V_{\min}$. In the subsequent chapters, circuit and architecture techniques used to mitigate SRAM variability and statistical CAD used in SRAM design will be discussed.

References

1. Y. Zorian, Embedded memory test and repair: infrastructure IP for SOC yield, in *Proceedings the International Test Conference (ITC)*, 2002, pp. 340–349
2. R. Kuppuswamy, S. Sawant, S. Balasubramanian, P. Kaushik, N. Natarajan, J. Gilbert, Over one million TPCC with a 45nm 6-core Xeon ® CPU, in *IEEE International Solid-State Circuits Conference—Digest of Technical Papers, ISSCC 2009*, Febraury 2009, pp. 70–71, 71a
3. H. Pilo, IEDM SRAM short course, (2006)
4. H. Yamauchi, Embedded SRAM circuit design technologies for a 45 nm and beyond, in: *ASICON '07: 7th International Conference on ASIC*, 22–25 October 2007, pp. 1028–1033
5. K. Agarwal, S. Nassif, Statistical analysis of SRAM cell stability, in *DAC '06: Proceedings of the 43rd Annual Conference on Design Automation*, 2006, pp. 57–62
6. A. Bhavnagarwala, S. Kosonocky, C. Radens, K. Stawiasz, R. Mann, Q. Ye, K. Chin, Fluctuation limits and scaling opportunities for CMOS SRAM cells, in *Proceedings of the International Electron Devices Meeting (IEDM)*, 2005, pp. 659–662
7. H. Pilo, C. Barwin, G. Braceras, C. Browning, S. Lamphier, F. Towler, An SRAM design in 65-nm technology node featuring read and write-assist circuits to expand operating voltage. IEEE J. Solid-State Circuits **42**(4), 813–819 (2007)
8. M. Yamaoka, N. Maeda, Y. Shinozaki, Y. Shimazaki, K. Nii, S. Shimada, K. Yanagisawa, T. Kawahara, 90-nm process-variation adaptive embedded SRAM modules with power-line-floating write technique. IEEE J. Solid-State Circuits **41**(3), 705–711 (2006)
9. S. Mukhopadhyay, H. Mahmoodi, K. Roy, Statistical design and optimization of SRAM cell for yield enhancement, in *Proceedings of International Conference on Computer Aided Design*, 2004, pp. 10–13
10. The International Technology Roadmap for Semiconductors (ITRS). http://public.itrs.net
11. M. Sinangil, H. Mair, A. Chandrakasan, A 28 nm high-density 6T SRAM with optimized peripheral-assist circuits for operation down to 0.6 V, in *IEEE International Solid-State Circuits Conference Digest of Technical Papers (ISSCC)*, Febraury 2011, pp. 260–262
12. B. Wong, A. Mittal, Y. Cao, G.W. Starr, *Nano-CMOS Circuit and Physical Design* (Wiley-Interscience, New York, 2004)
13. S. Sapatnekar, *Timing* (Springer, Boston, 2004)
14. A. Chandrakasan, W.J. Bowhill, F.Fox, *Design of High-Performance Microprocessor Circuits* (Wiley-IEEE Press, Piscataway, 2000)
15. A. Srivastava, D. Sylvester, D. Blaauw, *Statistical Analysis and Optimization for VLSI: Timing and Power (Series on Integrated Circuits and Systems)* (Springer, Boston, 2005)
16. Y. Cheng, C. Hu, *MOSFET Modeling and BSIM User Guide* (Kluwer Academic Publishers, Boston, 1999)
17. J. Tschanz, K. Bowman, V. De, Variation-tolerant circuits: circuit solutions and techniques, in *DAC '05: Proceedings of the 42nd Annual Conference on Design automation*, 2005, pp. 762–763
18. S. Borkar, T. Karnik, S. Narendra, J. Tschanz, A. Keshavarzi, V. De, Parameter variations and impact on circuits and microarchitecture, in *DAC '03: Proceedings of the 40th conference on Design automation*, 2003, pp. 338–342

19. T. Karnik, S. Borkar, V. De, Sub-90 nm technologies: challenges and opportunities for CAD, in *ICCAD '02: Proceedings of the IEEE/ACM International Conference on Computer-Aided Design*, 2002, pp. 203–206
20. S. Borkar, T. Karnik, V. De, Design and reliability challenges in nanometer technologies, in *DAC '04: Proceedings of the 41st Annual Conference on Design Automation*, 2004, pp. 75–75
21. A. Keshavarzi, G. Schrom, S. Tang, S. Ma, K. Bowman, S. Tyagi, K. Zhang, T. Linton, N. Hakim, S. Duvall, J. Brews, V. De, Measurements and modeling of intrinsic fluctuations in MOSFET threshold voltage, in *ISLPED '05: Proceedings of the 2005 International Symposium on Low Power Electronics and Design*, 2005, pp. 26–29
22. K. Bowman, S. Duvall, J. Meindl, Impact of die-to-die and within-die parameter fluctuations on the maximum clock frequency distribution for gigascale integration. IEEE J. Solid-State Circuits **37**(2), 183–190 (2002)
23. H. Masuda, S. Ohkawa, A. Kurokawa, M. Aoki, Challenge: variability characterization and modeling for 65- to 90-nm processes, in *Proceedings of IEEE Custom Integrated Circuits Conference*, 2005, pp. 593–599
24. J.A. Croon, W. Sansen, H.E. Maes, *Matching Properties of Deep Sub-Micron MOS Transistors* (Springer, New York, 2005)
25. Y. Taur, T.H. Ning, *Fundamentals of Modern VLSI Devices* (Cambridge University Press, New York, 1998)
26. T. Mizuno, J. Okumtura, A. Toriumi, Experimental study of threshold voltage fluctuation due to statistical variation of channel dopant number in MOSFET's. IEEE Trans. Electron Devices **41**(11), 2216–2221 (1994)
27. K. Takeuchi, T. Tatsumi, A. Furukawa, Channel engineering for the reduction of random-dopant-placement-induced threshold voltage fluctuations, in *Proceedings of the International Electron Devices Meeting (IEDM)*, 1996, pp. 841–844
28. A. Asenov, A. Brown, J. Davies, S. Kaya, G. Slavcheva, Simulation of intrinsic parameter fluctuations in decananometer and nanometer-scale MOSFETs. IEEE Trans. Electron Devices **50**(9), 1837–1852 (2003)
29. K. Sonoda, K. Ishikawa, T. Eimori, O. Tsuchiya, Discrete dopant effects on statistical variation of random telegraph signal magnitude. IEEE Trans. Electron Devices. **54**(8), 1918–1925 (2007)
30. M. Miyamura, T. Fukai, T. Ikezawa, R. Ueno, K. Takeuchi, M. Hane, SRAM critical yield evaluation based on comprehensive physical/statistical modeling, considering anomalous non-gaussian intrinsic transistor fluctuations, *Proceedings of IEEE Symposium on VLSI Technology*, June 2007, pp. 22–23
31. M. Pelgrom, A. Duinmaijer, A. Welbers, Matching properties of MOS transistors. IEEE J. Solid-State Circuits **24**(5), 1433–1439 (1989)
32. D. Frank, R. Dennard, E. Nowak, P. Solomon, Y. Taur, H.S. Wong, Device scaling limits of Si MOSFETs and their application dependencies. Proc. IEEE **89**(3), 259–288 (2001)
33. Y. Ye, F. Liu, M. Chen, S. Nassif, Y. Cao, Statistical modeling and simulation of threshold variation under random dopant fluctuations and line-edge roughness, IEEE Trans. Very Large Scale Integr. (VLSI) Syst. **19**(6), 987–996 (2011)
34. B. Razavi, *Design of Analog CMOS Integrated Circuits* (McGraw-Hill, New York, 2000)
35. T.-C. Chen, Where is CMOS going: trendy hype versus real technology, in *Proceedings of the International Solid-State Circuits Conference ISSCC*, 2006, pp. 22–28
36. K. Kuhn, CMOS transistor scaling past 32 nm and implications on variation, in *Advanced Semiconductor Manufacturing Conference (ASMC), IEEE/SEMI*, July 2010, pp. 241–246
37. J.-T. Kong, CAD for nanometer silicon design challenges and success. IEEE Trans. Very Large Scale Integr. Syst. **12**(11), 1132–1147 (2004)
38. P. Friedberg, Y. Cao, J. Cain, R. Wang, J. Rabaey, C. Spanos, Modeling within-die spatial correlation effects for process-design co-optimization, in *ISQED '05: Proceedings of the Sixth International Symposium on Quality of Electronic Design*, 2005, pp. 516–521
39. C. Wu, Y. Leung, C. Chang, M. Tsai, H. Huang, D. Lin, Y. Sheu, C. Hsieh, W. Liang, L. Han, W. Chen, S. Chang, S. Wu, S. Lin, H. Lin, C. Wang, P. Wang, T. Lee, C. Fu, C. Chang,

S. Chen, S. Jang, S. Shue, H. Lin, Y. See, Y. Mii, C. Diaz, B. Lin, M. Liang, Y. Sun, A 90-nm CMOS device technology with high-speed, general-purpose, and low-leakage transistors for system on chip applications, in *Proceedings of the International Electron Devices Meeting (IEDM)*, 2002, pp. 65–68

40. K. Takeuchi, T. Nagumo, K. Takeda, S. Asayama, S. Yokogawa, K. Imai, Y. Hayashi, Direct observation of RTN-induced SRAM failure by accelerated testing and its application to product reliability assessment, in *Symposium on VLSI Technology (VLSIT)*, June 2010, pp. 189–190
41. N. Tega, H. Miki, F. Pagette, D. Frank, A. Ray, M. Rooks, W. Haensch, K. Torii, Increasing threshold voltage variation due to random telegraph noise in FETs as gate lengths scale to 20 nm, in *Symposium on VLSI Technology*, June 2009, pp. 50–51
42. N. Tega, H. Miki, M. Yamaoka, H. Kume, T. Mine, T. Ishida, Y. Mori, R. Yamada, K. Torii, Impact of threshold voltage fluctuation due to random telegraph noise on scaled-down SRAM, in *IEEE International Reliability Physics Symposium IRPS 2008*, 27-May 1 2008, pp. 541–546
43. S. Bhardwaj, W. Wang, R. Vattikonda, Y. Cao, S. Vrudhula, Predictive modeling of the NBTI effect for reliable design, in *Custom Integrated Circuits Conference CICC '06, IEEE*, Septemper 2006, pp. 189–192
44. J. Lin, A. Oates, H. Tseng, Y. Liao, T. Chung, K. Huang, P. Tong, S. Yau, Y. Wang, Prediction and control of NBTI induced SRAM Vccmin drift, in *Proceedings of the International Electron Devices Meeting (IEDM)*, 2006
45. S. Chakravarthi, A. Krishnan, V. Reddy, C. Machala, S. Krishnan, A comprehensive framework for predictive modeling of negative bias temperature instability, in *Proceedings of the 42nd Annual IEEE International Reliability Physics Symposium*, April 2004, pp. 273–282
46. S. Drapatz, K. Hofmann, G. Georgakos, D. Schmitt-Landsiedel, Impact of fast-recovering NBTI degradation on stability of large-scale SRAM arrays, in *Proceedings of the European Solid-State Device Research Conference (ESSDERC)*, September 2010, pp. 146–149
47. J. Lin, A. Oates, C. Yu, Time dependent Vccmin degradation of SRAM fabricated with High-k gate dielectrics, in *proceedings of the IEEE International 45th Annual Reliability Physics Symposium*, April 2007, pp. 439–444
48. P. Kinget, Device mismatch and tradeoffs in the design of analog circuits. IEEE J. Solid-State Circuits **40**(6), 1212–1224 (2005)
49. E. Karl, Y. Wang, Y.-G. Ng, Z. Guo, F. Hamzaoglu, U. Bhattacharya, K. Zhang, K. Mistry, M. Bohr, A 4.6 GHz 162 Mb SRAM design in 22 nm tri-gate CMOS technology with integrated active VMIN-enhancing assist circuitry, in *IEEE International Solid-State Circuits Conference Digest of Technical Papers (ISSCC)*, Febraury 2012, pp. 230–232
50. S. Damaraju, V. George, S. Jahagirdar, T. Khondker, R. Milstrey, S. Sarkar, S. Siers, I. Stolero, A. Subbiah, A 22 nm IA multi-CPU and GPU system-on-chip, in *Solid-State Circuits Conference Digest of Technical Papers (ISSCC), IEEE International*, Febraury 2012, pp. 56–57
51. K. Kuhn, Reducing variation in advanced logic technologies: Approaches to process and design for manufacturability of nanoscale CMOS, in *IEEE International Electron Devices Meeting IEDM 2007*, December 2007, pp. 471–474
52. Y. Liu, K. Endo, S. O'uchi, T. Kamei, J. Tsukada, H. Yamauchi, Y. Ishikawa, T. Hayashida, K. Sakamoto, T. Matsukawa, A. Ogura, and M. Masahara, On the gate-stack origin threshold voltage variability in scaled FinFETs and multi-FinFETs, in *Symposium on VLSI Technology (VLSIT)*, June 2010, pp. 101–102
53. M. Khellah, D. Khalil, D. Somasekhar, Y. Ismail, T. Karnik, V. De, Effect of power supply noise on SRAM dynamic stability, *Proceedings of IEEE Symposium on VLSI Circuits*, June 2007, pp. 76–77
54. R. Heald, P. Wang, Variability in sub-100 nm SRAM designs, in *Proceedings of International Conference on Computer Aided Design*, 2004, pp. 347–352
55. M. Yamaoka, N. Maeda, Y. Shinozaki, Y. Shimazaki, K. Nii, S. Shimada, K. Yanagisawa, T. Kawahara, Low-power embedded SRAM modules with expanded margins for writing, in *Proceedings of the International Solid-State Circuits Conference ISSCC*, 2005, vol. 1, pp. 480–611

56. M. Yabuuchi, K. Nii, Y. Tsukamoto, S. Ohbayashi, S. Imaoka, H. Makino, Y. Yamagami, S. lshikura, T. Terano, T. Oashi, K. Hashimoto, A. Sebe, G. Okazaki, K. Satomi, H. Akamatsu, H. Shinohara, A 45 nm low-standby-power embedded SRAM with improved immunity against process and temperature variations, in *Proceedings of the International Solid-State Circuits Conference ISSCC*, 11–15 Febraury 2007, pp. 326–606
57. H. Qin, Y. Cao, D. Markovic, A. Vladimirescu, J. Rabaey, SRAM leakage suppression by minimizing standby supply voltage, in *Proceedings of the International Symposium on Quality of Electronic Design ISQED*, 2004, pp. 55–60
58. J. Wang, A. Singhee, R. Rutenbar, B. Calhoun, Statistical modeling for the minimum standby supply voltage of a full SRAM array, in *33rd European Solid State Circuits Conference ESSCIRC*, September 2007, pp. 400–403
59. M.H. Abu-Rahma, K. Chowdhury, J. Wang, Z. Chen, S.S. Yoon, M. Anis, A methodology for statistical estimation of read access yield in SRAMs, in *DAC '08: Proceedings of the 45th Conference on Design Automation*, 2008, pp. 205–210
60. M. Yamaoka, T. Kawahara, Operating-margin-improved SRAM with column-at-a-time body-bias control technique, in *33rd European Solid State Circuits Conference ESSCIRC*, 11–13 September 2007, pp. 396–399
61. E. Seevinck, F. List, J. Lohstroh, Static-noise margin analysis of MOS SRAM cells. IEEE J. Solid-State Circuits **22**(5), 748–754 (1987)
62. K. Takeda, H. Ikeda, Y. Hagihara, M. Nomura, H. Kobatake, Redefinition of write margin for next-generation SRAM and write-margin monitoring circuit, in *Proceedings of the International Solid-State Circuits Conference ISSCC*, 2006, pp. 2602–2611
63. J.M. Rabaey, A. Chandrakasan, B. Nikolic, *Digital Integrated Circuits*, 2nd Edn. (Prentice Hall, Englewood Cliffs, 2002)
64. K. Aadithya, S. Venogopalan, A. Demir, J. Roychowdhury, Mustard: a coupled, stochastic/deterministic, discrete/continuous technique for predicting the impact of random telegraph noise on SRAMs and DRAMs, in *Design Automation Conference (DAC), 2011 48th ACM/EDAC/IEEE*, June 2011, pp. 292–297
65. F. shi Lai, C.-F. Lee, On-chip voltage down converter to improve SRAM read/write margin and static power for sub-nano CMOS technology, IEEE J. Solid-State Circuits **42**(9), 2061–2070 (2007)
66. A. Carlson, Mechanism of increase in SRAM Vmin due to negative-bias temperature instability, IEEE Trans. Device Mater. Reliab. **7**(3), 473–478 (2007)
67. A. Krishnan, V. Reddy, D. Aldrich, J. Raval, K. Christensen, J. Rosal, C. O'Brien, R. Khamankar, A. Marshall, W.-K. Loh, R. McKee, S. Krishnan, SRAM cell static noise margin and VMIN sensitivity to transistor degradation, in *Proceedings of the International Electron Devices Meeting (IEDM)*, 2006
68. G. La Rosa, W.L. Ng, S. Rauch, R. Wong, J. Sudijono, Impact of NBTI induced statistical variation to SRAM cell stability, in *Proceedings of the 44th Annual IEEE International Reliability Physics Symposium*, March 2006, pp. 274–282
69. K. Kang, H. Kufluoglu, K. Roy, M. Ashraful Alam, Impact of negative-bias temperature instability in nanoscale SRAM array: Modeling and analysis. IEEE Trans. Comput.-Aided Des. Integr. Circuits Syst. **26**(10), 1770–1781 (2007)
70. M. Ball, J. Rosal, R. McKee, W. Loh, T. Houston, R. Garcia, J. Raval, D. Li, R. Hollingsworth, R. Gury, R. Eklund, J. Vaccani, B. Castellano, F. Piacibello, S. Ashburn, A. Tsao, A. Krishnan, J. Ondrusek, T. Anderson, A screening methodology for VMIN drift in SRAM arrays with application to sub-65 nm nodes, in *International Electron Devices Meeting, IEDM '06*, December 2006, pp. 1–4
71. M. Agostinelli, J. Hicks, J. Xu, B. Woolery, K. Mistry, K. Zhang, S. Jacobs, J. Jopling, W. Yang, B. Lee, T. Raz, M. Mehalel, P. Kolar, Y. Wang, J. Sandford, D. Pivin, C. Peterson, M. DiBattista, S. Pae, M. Jones, S. Johnson, G. Subramanian, Erratic fluctuations of SRAM cache vmin at the 90 nm process technology node, in *Electron Devices Meeting IEDM Technical Digest. IEEE International*, December 2005, pp. 655–658

72. M. Tanizawa, S. Ohbayashi, T. Okagaki, K. Sonoda, K. Eikyu, Y. Hirano, K. Ishikawa, O. Tsuchiya, Y. Inoue, Application of a statistical compact model for random telegraph noise to scaled-SRAM Vmin analysis, in *Symposium on VLSI Technology (VLSIT)*, June 2010, pp. 95–96
73. S. O. Toh, T.-J. K. Liu, B. Nikolic, Impact of random telegraph signaling noise on SRAM stability, in *Symposium on VLSI Technology (VLSIT)*, June 2011, pp. 204–205
74. M. Yamaoka, H. Miki, A. Bansal, S. Wu, D. Frank, E. Leobandung, K. Torii, Evaluation methodology for random telegraph noise effects in SRAM arrays, in *IEEE International Electron Devices Meeting (IEDM)*, December 2011, pp. 32.2.1–32.2.4
75. S.O. Toh, Y. Tsukamoto, Z. Guo, L. Jones, T.-J.K. Liu, B. Nikolic, Impact of random telegraph signals on Vmin in 45 nm SRAM, in *IEEE International Electron Devices Meeting (IEDM)*, December 2009, pp. 1–4
76. K. Takeuchi, T. Nagumo, T. Hase, Comprehensive SRAM design methodology for RTN reliability, in *Symposium on VLSI Technology (VLSIT)*, June 2011, pp. 130–131
77. Y. Tsukamoto, S. O. Toh, C. Shin, A. Mairena, T.-J. K. Liu, B. Nikolić, Analysis of the relationship between random telegraph signal and negative bias temperature instability, in *IEEE International Reliability Physics Symposium (IRPS)*, May 2010, pp. 1117–1121
78. R. Baumann, The impact of technology scaling on soft error rate performance and limits to the efficacy of error correction, in *Proceedings of the International Electron Devices Meeting (IEDM)*, 2002, pp. 329–332
79. R. Baumann, Soft errors in advanced computer systems, IEEE Des. Test of Comput. **22**(3) 258–266 (2005)
80. E. Cannon, D. Reinhardt, M. Gordon, P. Makowenskyj, SRAM SER in 90, 130 and 180 nm bulk and SOI technologies, in *Proceedings of the 42nd IEEE International Reliability Physics Symposium*, 2004, pp. 300–304
81. A. Balasubramanian, P. Fleming, B. Bhuva, A. Sternberg, L. Massengill, Implications of dopant-fluctuation-induced v_t variations on the radiation hardness of deep submicrometer CMOS SRAMs. IEEE Trans. Device and Mater. Reliab. **8**(1), 135–144 (2008)
82. E. Cannon, A. KleinOsowski, R. Kanj, D. Reinhardt, R. Joshi, The impact of aging effects and manufacturing variation on SRAM soft-error rate, IEEE Trans. Device Mater. Reliab. **8**(1), 145–152 (2008)
83. S.S. Mukherjee, J. Emer, S.K. Reinhardt, The soft error problem: an architectural perspective, in *HPCA '05: Proceedings of the 11th International Symposium on High-Performance Computer Architecture*, 2005, pp. 243–247
84. E. Ibe, H. Taniguchi, Y. Yahagi, K.-i. Shimbo, and T. Toba, Impact of scaling on neutron-induced soft error in SRAMs from a 250 nm to a 22 nm design rule, IEEE Trans. Electron Devices **57**(7), 1527–1538 (2010)
85. A. Dixit, A. Wood, The impact of new technology on soft error rates, in *Reliability Physics Symposium (IRPS), 2011 IEEE International*, April 2011, pp. 5B.4.1–5B.4.7
86. N.H. Weste, D. Harris, *CMOS VLSI Design : A Circuits and Systems Perspective*, 3rd Edn. (Addison Wesley, Boston, 2004)
87. K. Itoh, M. Horiguchi, M. Yamaoka, Low-voltage limitations of memory-rich nano-scale CMOS LSIs, in *37th European Solid State Device Research Conference, ESSDERC 2007*, Septemper 2007, pp. 68–75
88. R. Rajsuman, Design and test of large embedded memories: an overview. IEEE Des. Test Comput. **18**(3), 16–27 (2001)
89. Y. Zorian, S. Shoukourian, Embedded-memory test and repair: infrastructure IP for SoC yield. IEEE Des. Test Comput. **20**(3), 58–66 (2003)
90. T. Chen, V.-K. Kim, M. Tegethoff, IC yield estimation at early stages of the design cycle. Microelectron. J. **30**(8), 725–732 (1999)
91. T. Barnett, M. Grady, K. Purdy, A. Singh, Redundancy implications for early-life reliability: experimental verification of an integrated yield-reliability model, in *Proceedings of the International Test Conference*, 2002, pp. 693–699

92. K. Roy, S. Mukhopadhyay, H. Mahmoodi-Meimand, Leakage current mechanisms and leakage reduction techniques in deep-submicrometer CMOS circuits. Proc. IEEE **91**(2), 305–327 (2003)
93. Y.-S. Lin, C.-C. Wu, C.-S. Chang, R.-P. Yang, W.-M. Chen, J.-J. Liaw, C. Diaz, Leakage scaling in deep submicron CMOS for SoC. IEEE Trans. Electron Devices **49**(6), 1034–1041 (2002)
94. J. Kao, S. Narendra, A. Chandrakasan, Subthreshold leakage modeling and reduction techniques, in *ICCAD '02: Proceedings of the IEEE/ACM International Conference on Computer-Aided Design*, 2002, pp. 141–148
95. S. Narendra, V. De, S. Borkar, D. Antoniadis, A. Chandrakasan, Full-chip subthreshold leakage power prediction and reduction techniques for sub-0.18 μm CMOS. IEEE J. Solid-State Circuits **39**(2), 501–510 (2004)
96. J.P. de Gyvez, H. Tuinhout, Threshold voltage mismatch and intra-die leakage current in digital CMOS circuits. IEEE J. Solid-State Circuits **39**(1), 157–168 (2004)
97. S. Narendra, V. De, S. Borkar, D. Antoniadis, A. Chandrakasan, Full-chip sub-threshold leakage power prediction model for sub-0.18 μm CMOS, in *ISLPED '02: Proceedings of the 2002 International Symposium on Low Power Electronics and Design*, 2002, pp. 19–23
98. R. Rao, A. Srivastava, D. Blaauw, D. Sylvester, Statistical analysis of subthreshold leakage current for VLSI circuits. IEEE Trans Very Large Scale Integr. Syst. **12**(2), 131–139 (2004)
99. M. Miyazaki, G. Ono, T. Hattori, K. Shiozawa, K. Uchiyama, K. Ishibashi, A 1000-MIPS/w microprocessor using speed adaptive threshold-voltage CMOS with forward bias, in *Proceedings of the International Solid-State Circuits Conference ISSCC*, 2000, pp. 420–421
100. M. Miyazaki, G. Ono, K. Ishibashi, A 1.2-GIPS/w microprocessor using speed-adaptive threshold-voltage CMOS with forward bias. IEEE J. Solid-State Circuits **37**(2), 210–217 (2002)
101. J. Tschanz, J. Kao, S. Narendra, R. Nair, D. Antoniadis, A. Chandrakasan, V. De, Adaptive body bias for reducing impacts of die-to-die and within-die parameter variations on microprocessor frequency and leakage. IEEE J. Solid-State Circuits **37**(11), 1396–1402 (2002)
102. J. Tschanz, S. Narendra, R. Nair, V. De, Effectiveness of adaptive supply voltage and body bias for reducing impact of parameter variations in low power and high performance microprocessors. IEEE J. Solid-State Circuits **38**(5), 826–829 (2003)
103. Y. Komatsu, K. Ishibashi, M. Yamamoto, T. Tsukada, K. Shimazaki, M. Fukazawa, M. Nagata, Substrate-noise and random-fluctuations reduction with self-adjusted forward body bias, in *Proceedings of IEEE Custom Integrated Circuits conference*, 2005, pp. 35–38
104. C. Kim, K. Roy, S. Hsu, A. Alvandpour, R. Krishnamurthy, S. Borkar, A process variation compensating technique for sub-90 nm dynamic circuits, in *Proceedings of IEEE Symposium on VLSI Circuits*, 2003, pp. 205–206
105. C. Kim, S. Hsu, R. Krishnamurthy, S. Borkar, K. Roy, Self calibrating circuit design for variation tolerant VLSI systems, in *IOLTS 2005 11th IEEE International On-Line Testing Symposium*, 2005, pp. 100–105
106. K. Bowman, S. Duvall, J. Meindl, Impact of die-to-die and within-die parameter fluctuations on the maximum clock frequency distribution, in *Proceedings of the International Solid-State Circuits Conference ISSCC*, 2001, pp. 278–279
107. K. Bowman, J. Meindl, Impact of within-die parameter fluctuations on future maximum clock frequency distributions, in *Proceedings of IEEE Custom Integrated Circuits Conference*, 2001, pp. 229–232
108. D. Marculescu, E. Talpes, Variability and energy awareness: a microarchitecture-level perspective, in *DAC '05: Proceedings of the 42nd Annual Conference on Design Automation*, 2005, pp. 11–16
109. D. Marculescu, E. Talpes, Energy awareness and uncertainty in microarchitecture-level design. IEEE Micro. **25**(5), 64–76 (2005)
110. A. Datta, S. Bhunia, S. Mukhopadhyay, N. Banerjee, K. Roy, Statistical modeling of pipeline delay and design of pipeline under process variation to enhance yield in sub-100 nm technologies, in *DATE '05: Proceedings of the Conference on Design, Automation and Test in Europe*, 2005, pp. 926–931

111. N. Azizi, M.M. Khellah, V. De, F.N. Najm, Variations-aware low-power design with voltage scaling, in *DAC '05: Proceedings of the 42nd Annual Conference on Design Automation*, 2005, pp. 529–534
112. E. Humenay, D. Tarjan, K. Skadron, Impact of parameter variations on multi-core chips, in *Proceedings of the 2006 Workshop on Architectural Support for Gigascale Integration, in Conjunction with the 33rd International Symposium on Computer Architecture (ISCA)*, June 2006
113. E. Humenay, W. Huang, M.R. Stan, K. Skadron, Toward an architectural treatment of parameter variations, University of Virginia, Department of Computer Science, Technical Report. CS-2005-16, September 2005
114. S.H. Choi, B.C. Paul, K. Roy, Novel sizing algorithm for yield improvement under process variation in nanometer technology, in *DAC '04: Proceedings of the 41st Annual Conference on Design Automation*, 2004, pp. 454–459
115. A. Agarwal, K. Chopra, D. Blaauw, Statistical timing based optimization using gate sizing, in *DATE '05: Proceedings of the Conference on Design, Automation and Test in, Europe*, 2005, pp. 400–405
116. F.N. Najm, On the need for statistical timing analysis, in *DAC '05: Proceedings of the 42nd Annual Conference on Design Automation*, 2005, pp. 764–765

Chapter 3
Variation-Tolerant SRAM Write and Read Assist Techniques

3.1 Introduction

There are stringent requirements to lower the power consumption and achieve higher speed in today's SoCs and microprocessors. Voltage scaling combined with technology scaling has been effective in achieving both requirements. However, the large increase in random variations in advanced CMOS technology nodes has created several challenges for SRAM design. This is exacerbated by the high demand for low voltage and high density memories for SoC [1]. Dealing with SRAM cell stability at lower supply voltages is currently one of the biggest challenges in SRAM design [2–4].

3.2 SRAM Stability Metrics

Several metrics have been proposed to analyze the SRAM stability margins. These metrics are beneficial in studying how different design options affect SRAM stability, analyzing the effectiveness of assist techniques, predicting yield, optimizing the memory V_{min}, as well as determining the process window at low voltage, as shown in Fig. 3.1. At low voltage, the write and read margins reduce, which diminishes the operation window. In the next section, we study different metrics used for write and read margins.

Static metrics are typically easier to simulate and measure since they assume that bitcell stability is time-independent. Static metrics have been used extensively in measurement and simulation of bitcell stability [3, 4, 6–11].

Dynamic stability metrics are time-dependent, and have been proposed to improve the accuracy of determining SRAM stability. However, dynamic metrics are more difficult to simulate and analyze since they require transient simulation, while static

M. H. Abu-Rahma and M. Anis, *Nanometer Variation-Tolerant SRAM*,

DOI: 10.1007/978-1-4614-1749-1_3,

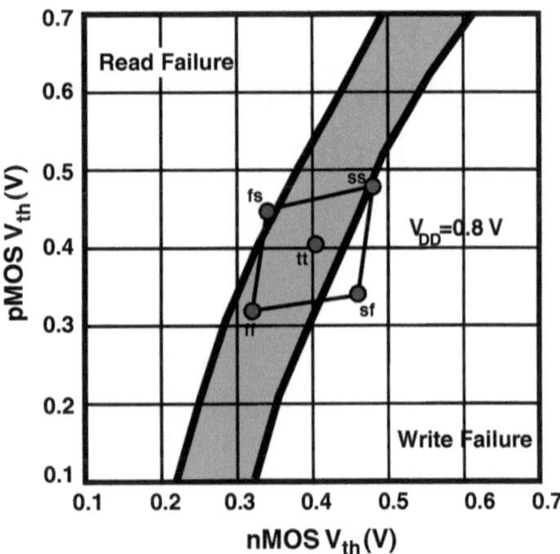

Fig. 3.1 Write and read stability metrics used to determine the V_{th} window of operation [5]

metrics may be determined using DC simulation. In the next sections, we review the commonly used static and dynamic metrics for SRAM write and read stability [3, 4, 6–11].

3.2.1 Static Write Margin

In write operation, BLB is pulled to zero using write driver, while WL is enabled, as shown in Fig. 3.2. Therefore, the NMOS PG2 is turned ON, which results in a voltage drop in the storage node QB holding data 1. When this voltage falls below $V_{DD} - V_{th}$ for the PU1, PU1 starts the feedback action. For stable write operation, PG2 should be stronger than PU2.

To quantify write stability, one of the most widely used metrics is the write static noise margin (WSNM), which uses a similar concept to the read SNM described in Sect. 3.2.3. In this approach, the voltage transfer characteristics (VTC) of the two sides of the bitcell are obtained from DC simulation, while the bitlines (BL and BLB) are driven to the write operation condition (i.e., BL connected to V_{DD} and BLB to zero, as shown in Fig. 3.3). For a successful write operation, there should only be one cross-point between the VTCs of the bitcell inverters, which implies that there is only one solution (the cell is monostable), as shown in Fig. 3.3a. The separation between the two VTCs indicates the bitcell write margin. If the separation is reduced, the bitcell write ability worsens until the separation reaches zero, which means that there are two or three cross points, so the bitcell changes from monostable to bistable or metastable. Therefore, to measure write stability, the WSNM is defined as the width of the minimum square that can be enclosed between the two VTCs. Using this definition, the bitcell failure condition is defined as WSNM =0 [12]. Improvements in the definition of WSNM have been proposed [6].

Another write stability metric, the bitline write margin (WVBL), is derived from the BL voltage, as shown in Fig. 3.3b. In this method, the bitcell is configured in write operation with one bitline (BL) connected to V_{DD}, and the other bitline (BLB)

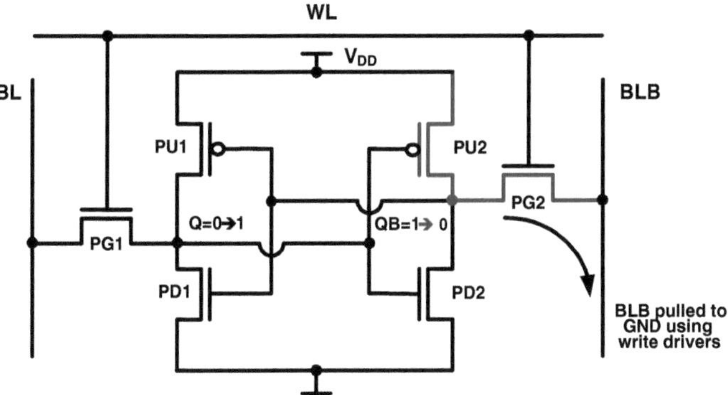

Fig. 3.2 Bitcell in write operation

swept from V_{DD} to zero. As the BLB voltage is reduced, the voltage at which the internal nodes (Q and QB) flip is defined as the bitline write margin (WVBL). The lower the bitline, the more difficult it is to write the bitcell. WVBL can also be defined as the maximum bitline voltage at which the bitcell can be written. The write failure criteria in this case is defined as WVBL =0 [12].

Similar to the WVBL write margin definition, the WL voltage can be used to assess write ability. In this case, the WL voltage is swept from zero to V_{DD} to find the minimum voltage at which the bitcell is written (Q and QB flip) and is called $WVWL$, as shown in Fig. 3.3c. The lower the value of WVWL, the easier it is to write on the bitcell. The write failure criteria is defined as WVWL $\geq V_{DD}$ [12].

Other DC write margins have been proposed, such as the N-curve method, as shown in Fig. 3.3d, which uses current information instead of voltage [4, 12].

It is important to note that due to device variations, the bitcell margin needs to be evaluated on both sides of the bitcell (i.e., in the case of writing zero via BL or via BLB), since variations will cause the margin to be different for each side of the cell. Hence, the write margin becomes the worst of the two sides margins.

3.2.2 Dynamic Write Margin

The main limitation of DC metrics is that they ignore the nonlinear dynamics of write operation by neglecting the bitcell time-dependence. For example, static metrics assume that the WL pulse width is infinite and only account for the voltage amplitude. In reality, the write pulse width is proportional to the memory cycle time. In addition, static margins assume that one bitline is actively pulled to zero while the other bitline stays at V_{DD}, which is not realistic in dynamic operation.

Figure 3.4 shows the distribution of bitcell write time, defined as the time required for the internal storage node to flip after the WL is enabled. The distribution is

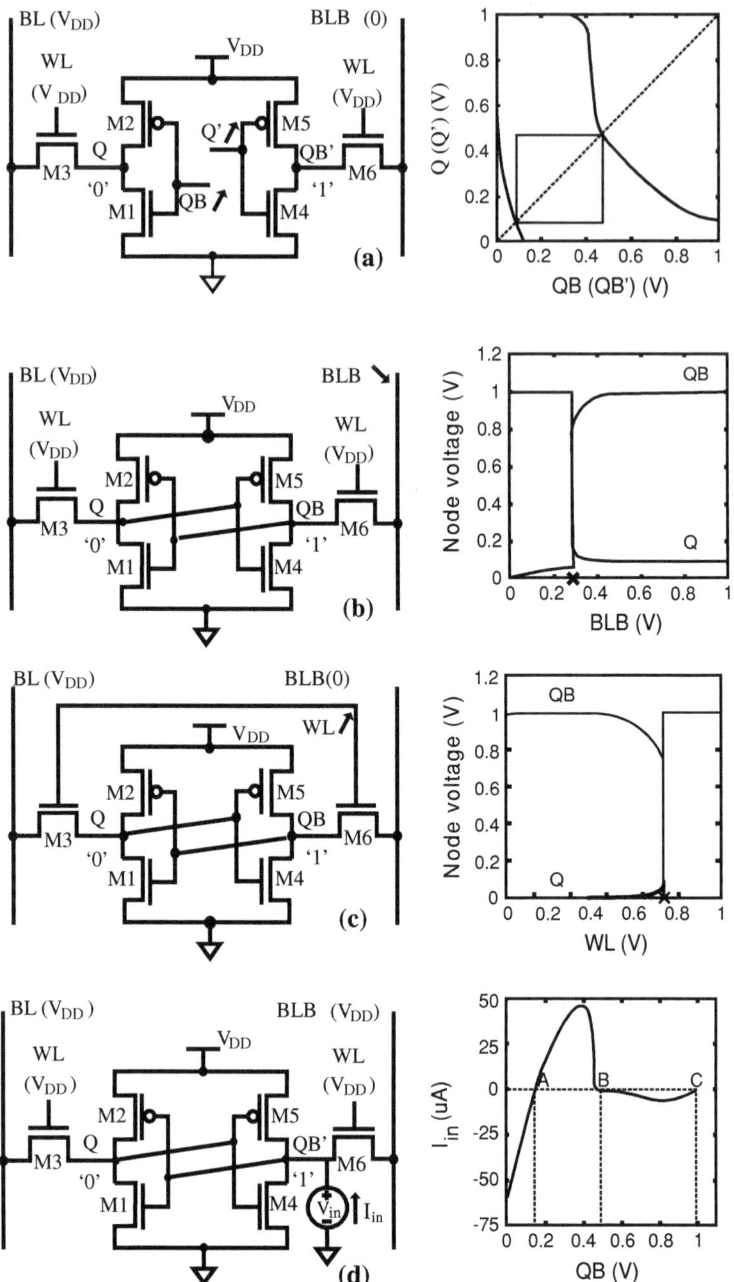

Fig. 3.3 Different DC definitions for bitcell write margin including **a** write SNM, **b** WVBL, **c** WVWL, and **d** N-curve [12]

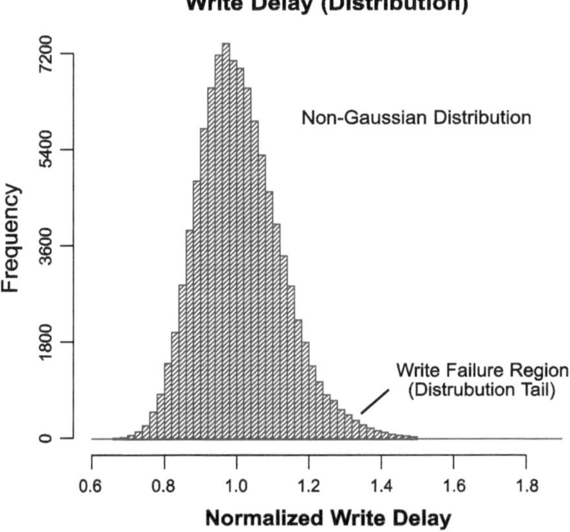

Fig. 3.4 Distribution of bitcell write time which shows that the distribution is non-Gaussian with a long tail toward higher write time [13]

not a Gaussian distribution, and shows a long tail toward higher bitcell write time. Dynamic write failure can be defined as the probability that the bitcell write time is larger than the WL pulse width [13–15]. Other definitions use the differential voltage between bitcells storage nodes at the end of write operation, as shown in Fig. 3.5. The impact of WL pulse width on write dynamics is shown in Fig. 3.6. As the WL pulse width increases, the failure probability (or measured fail bit count) decreases and approaches the DC value.

Comparisons between static and dynamic metrics have shown that static metrics are optimistic when predicting write failure probability [10, 13, 16], in some cases by 3 orders of magnitude [10]. This underestimation of failure results from static metrics' assumption that the WL pulse width is infinite; therefore, failures due to insufficient WL pulse width are not captured in the static write margin. This causes dynamic write V_{min} to be typically larger than static V_{min} [15]. Moreover, the fact that the distribution of dynamic write margin is not Gaussian complicates the statistical yield estimation [12].

One of the interesting findings from dynamic write stability is the relationship between PMOS pull-up strength and write fails. Adjusting the PMOS pull-up to be weaker improves static write margin; however, it may reduce dynamic stability since the bitcell write time (time to flip) increases due to slower rising time of the bitcell storage node [6, 19]. As shown in Fig. 3.7, this type of dynamic failure causes frequency-dependent failure since the bitcell internal node stays near the metastable point and may take several cycles to reach V_{DD}. Hence, read operation fails if it is performed immediately after write operation [6, 19].

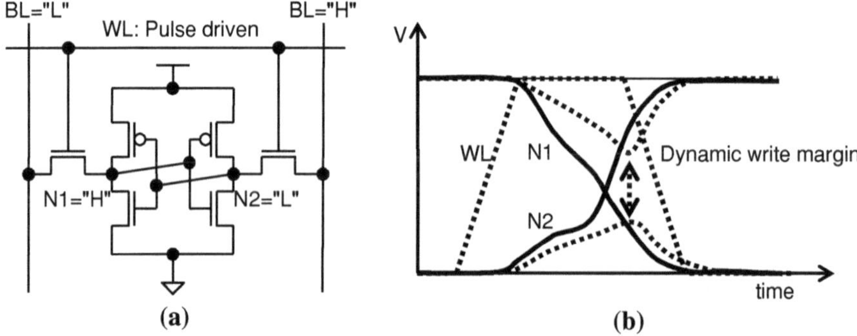

Fig. 3.5 Dynamic write margin definition based on transient analysis. **a** Memory cell state during dynamic (transient) analysis **b** Waveform during dynamic analysis [17]

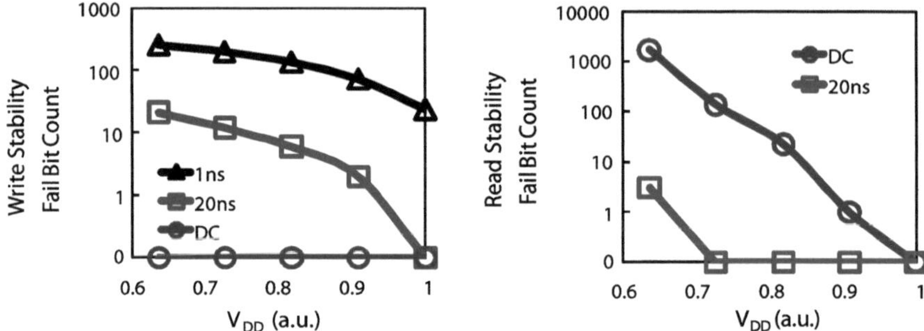

Fig. 3.6 Measured fail bit count dependence on V_{DD} and wordline pulse width for write and read stability [18]

3.2.3 Static Read Stability

In read operation, read disturb may occur after the wordline is enabled. The voltage at the internal storage node storing a zero (Q) slightly rises due to the voltage divider between the pass-gate transistor (PG1) and the pull-down (PD1), as shown in Fig. 3.8. If the voltage at Q rises above the threshold voltage of PD2, the cell may flip its state. In this case, stable read operation requires that PD1 should be stronger than PG1. Read stability failure increases with process variations, which affect all the transistors in the bitcell [3, 13, 20].

To quantify the bitcell's robustness against this type of failure, static noise margin (SNM) is one of the most commonly used metrics [21]. SNM is defined as the maximum amount of voltage noise that a cell can tolerate [21]. SNM is calculated by finding the largest square which fits inside the VTCs (butterfly curves), as shown in Fig. 3.9. A larger SNM implies higher robustness for the bitcell. However, due to WID variations, each transistor in the bitcell experiences different magnitude of variation, hence, the symmetry of the bitcell is lost. This causes large spread in SNM

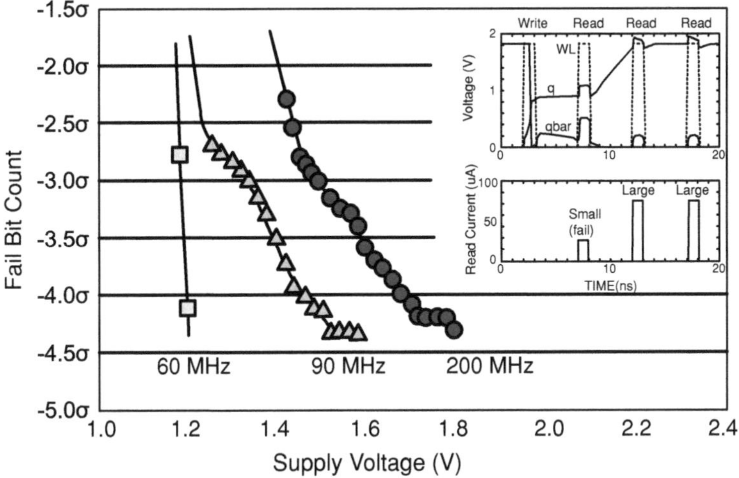

Fig. 3.7 Measured frequency-dependent failure attributed to dynamic write fail. The inset shows simulation of the write failure where the bitcell internal node (V_{n1}) does not fully reach V_{DD} after write operation in one cycle. The failure appears as a frequency-dependent failure since at lower frequency V_{n1} has sufficient time to reach V_{DD} before the next cycle [19]

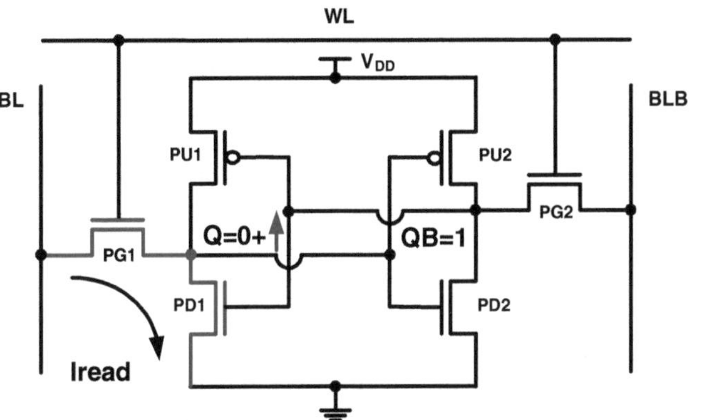

Fig. 3.8 Bitcell in read operation

as shown in measured SNM butterfly curves in Fig. 2.29. Read stability failure occurs if SNM reaches zero [13, 20, 21].

Other techniques for static read stability margins use the bitcell current such as the N-curves method [4, 23]. Because SNM is estimated using DC simulation, this method of measuring read stability does not account for the dynamic nature of read operation. In the next section, we present metrics for dynamic read stability.

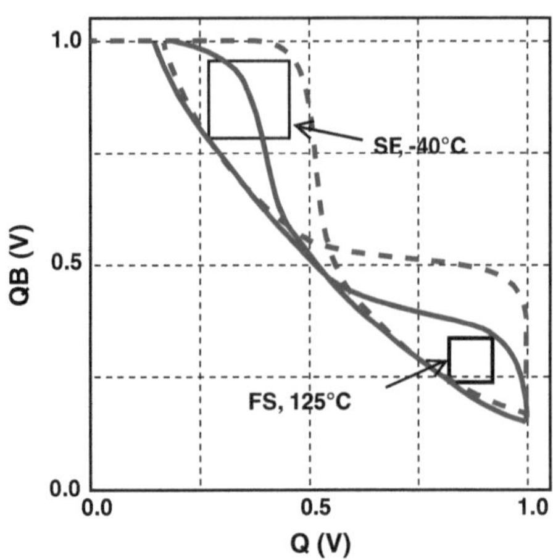

Fig. 3.9 SNM butterfly curves for a 45 nm SRAM bitcell simulated at skewed process corners, FS: fast NMOS, slow PMOS and SF: slow NMOS, fast PMOS. In this simulation, WID variations are not included, so the curves are symmetrical [22]

3.2.4 Dynamic Read Stability

Conventional SNM read stability metric cannot capture the dynamic behavior of the WL, BL, and internal storage nodes of the bitcell. Several works have recently investigated how the dynamic operation of the bitcell affects read stability. DC SNM gives pessimistic results compared to dynamic stability because DC SNM assumes the WL pulse width is infinite and that the BL is actively pulled to V_{DD} in read operation. In reality, the WL pulse width is typically a percentage of the cycle time, hence, the bitcell disturb may not have enough time to flip within the WL pulse width. Also, BLs are typically not pulled high in read operation, but are precharged before the WL is enabled, therefore, the BL capacitance has strong impact on the dynamic stability of the bitcell. Conventional static SNM can overestimate the probability of read flip failure by 6 orders of magnitude [10]. Moreover, static metrics cannot account for dynamic issues such as the impact of supply noise on stability [10].

Several dynamic read margin definitions have been proposed [10, 17, 24, 25]. Figure 3.10 shows the definition of dynamic read stability based on transient simulation [17]. The margin is defined as the minimum differential voltage between the storage nodes when the WL is high. The dynamic read margin strongly depends on the WL pulse width and the bitline capacitance [17, 26].

Several models have been developed to study the dynamic stability. Rigorous analysis using nonlinear system theory has shown that two conditions are required to cause a bitcell flip: the noise should, 1- exceed the static SNM, and 2- be sustained longer than a minimum critical duration [9, 25, 27, 28].

The impact of WL pulse width on dynamic read stability is shown in Fig. 3.6. As the WL pulse width decreases, the read flip failure probability (or measured fail bit count) decreases drastically. The impact of bitline capacitance on read failure can be estimated by simulating dynamic read stability, and shows that reducing the BL capacitance results into significant read stability improvement. Figure 3.11 shows

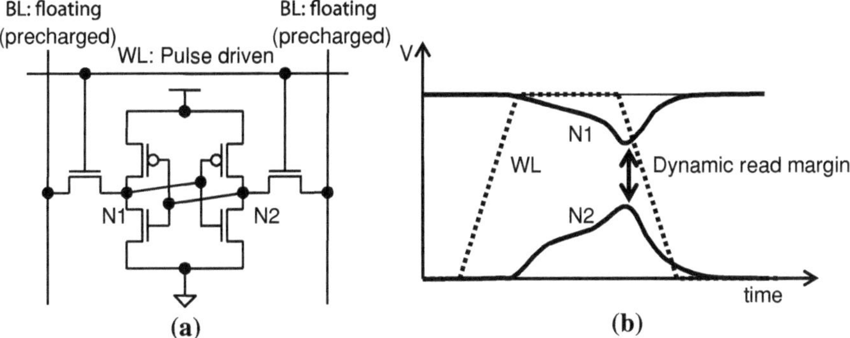

Fig. 3.10 Dynamic read margin definition based on transient analysis with the bitlines floating [17] **a** Memory cell state during dynamic(transient) analysis **b** Waveform during analysis

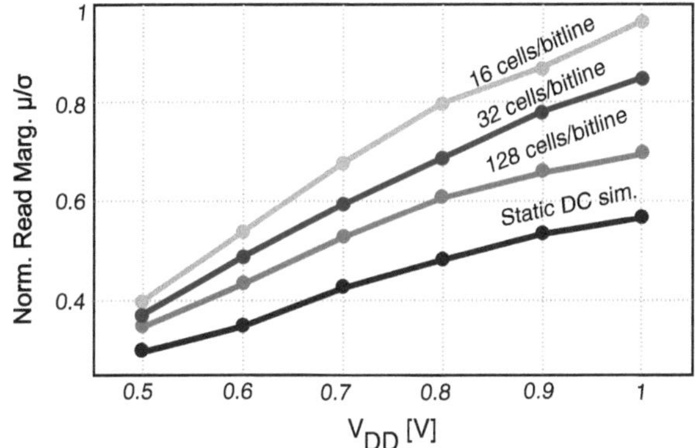

Fig. 3.11 Simulation of static and dynamic read stability margins for a 6T bitcell in 28 nm technology. Dynamic margins show much lower $V_{\min}$ compared to static [26]

simulation results for static and dynamic read margins for 6T cell in 28 nm technology. For the same read margin (constant μ/σ), $V_{\min}$ calculated by dynamic read margin is much lower $V_{\min}$ than that predicted by static simulation (0.7 V for dynamic while 0.9 V for static). Moreover, reducing the bitline capacitance by reducing the number of bitcells per bitline, can be used to further reduce by 100 mV when using 16 cells/bitline instead of 128 cells/bitline. This allows the bitcell to operate at lower voltage at the expense of larger area overheard, since additional read circuitry is required when using shorter bitlines (larger number of memory banks) [10, 24, 26, 29]. The dynamic read stability concept has wide applications in the area of read assist techniques, as will be discussed in the next sections.

Table 3.1 Impact of bitcell size increase on V_{min} reduction [30]

Bitcell area (normalized)	V_{min} (V)
1	1.1
1.11	1.0
1.22	0.8
1.44	0.7

3.3 Bitcell Design for Low Voltage Operation

The bitcell V_{min} depends strongly on random variations. To reduce SRAM V_{min}, the simplest technique is to increase the 6T bitcell area. Since the V_{th} variations are inversely proportional to the square root of device area, increasing the device area reduces the variations, which helps improve the bitcell stability, as shown in Table 3.1. However, increasing bitcell size comes with large area overhead, which increases the product cost significantly and limits the usage of large area 6T bitcells to small density memories.

Due to the conflicting requirements of read and write stability, it is challenging to reduce the V_{min} for 6T bitcell without a large increase in bitcell area. Instead of 6T, an 8T bitcell has been proposed to lower V_{min} by decoupling the read and write operations. Figure 3.12 shows an 8T bitcell, which has isolated read and write ports. The read operation is single-ended, using the read bitline (RBL) and read wordline (RWL), which eliminates read disturb failure since there is no contention between the pass-gate and the pull-down. The write operation is performed by differential write using WBL, $WBLB$, and the wordline (WL) which is similar to that in 6T bitcell. Since the WWL, WBL, and $WBLB$ are not used in read operation, bitcell devices they can be optimized solely for write operation where PG/PU current ratio can be increased without considering the PG/PD read constraint in 6T bitcell. Since the 8T bitcell has a dedicated write port, it does not suffer from read disturb and the limit for low voltage read stability is retaining the data (retention failure). Figure 3.13 compares the SNM for 6T and 8T bitcells in 32 nm technology node. The SNM for the 8T bitcell is 76 % higher than the 6T bitcell which allows the 8T bitcell to have lower V_{min} than the 6T.

While the 8T bitcell provides lower V_{min}, it adds 30–40 % area penalty [26, 31], as shown in Fig. 3.14. Also, 8T read operation is single-ended, which requires large signal sensing and limits the maximum number of cells on the bitline. The other option for read operation is to use a sense amplifier with one input connected to a reference voltage, which is typically slower and more susceptible to supply noise compared to differential sensing. Another difference between 8T and 6T bitcells is that the 8T bitcell does not allow column muxing, which is also called column interleaving. Therefore, the periphery circuits for 8T bitcells are larger than the 6T since each column requires dedicated read and write circuitry and cannot be shared for multiple columns. Moreover, without column interleaving, 8T bitcells become

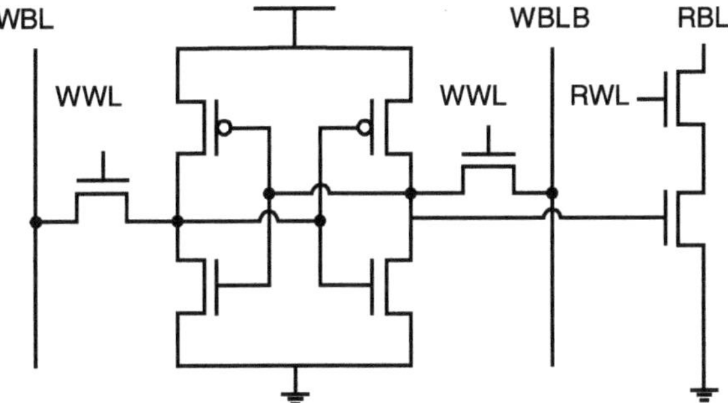

Fig. 3.12 8T bitcell with separate read and write ports which provides lower $V_{\min}$ compared to conventional 6T bitcell [31]

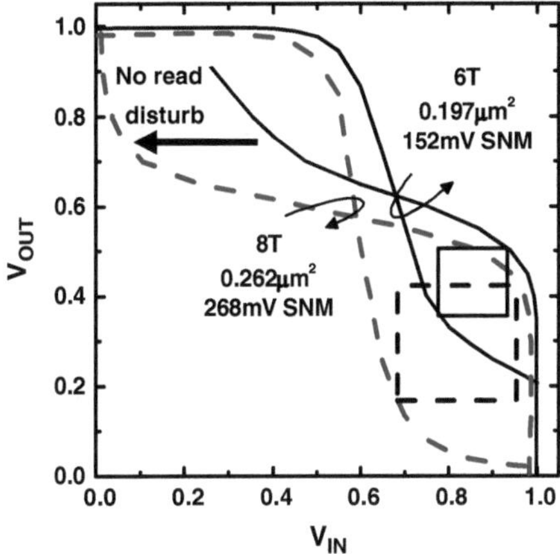

Fig. 3.13 8T bitcell provides higher SNM compared to 6T due to elimination of read disturbs [31]

more sensitive to soft errors, where multi-cell upsets (MCUs) cause multiple fails in the same word and may not be corrected using ECC, as discussed in Chap. 2.

In efforts to achieve the benefits of an 8T cell while reducing the area overhead, a 7T bitcell has been proposed as shown in Fig. 3.15. An NMOS transistor (N5) is added between the storage node (V2) and the pull-down device (N2) [32]. The WWL is used for write operation while WL is used for both read and write operations. The protection device N5 is controlled using the complement of WL ($/WL$). In read operation, WL is high and $/WL$ is low, hence, $N5$ prevents $V2$ from decreasing and the cell cannot flip even if V1 increases. In write operation, both WL and WWL are high, and one of the bitlines is pulled to zero, similar to 6T bitcell write operation. Read operation is single-ended using only BL.

A major disadvantage of 7T bitcells is that if $V2$ is storing a zero, then the node is floating, which requires a dynamic retention condition. If the WL activation period is longer than that the data retention time of $V2 = 0$, the stored data will be lost due

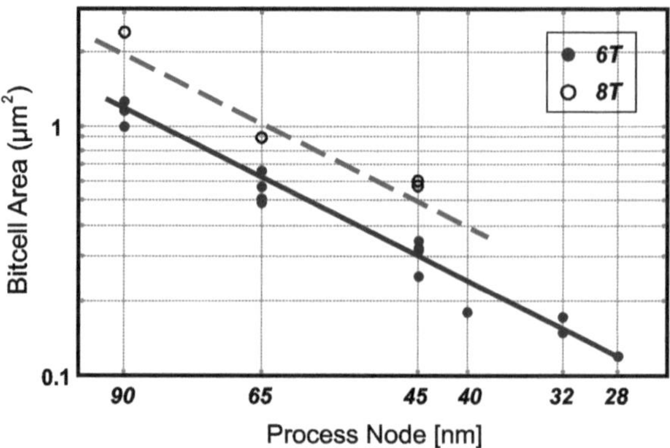

Fig. 3.14 Bitcell area scaling for 6T and 8T bitcells [26]

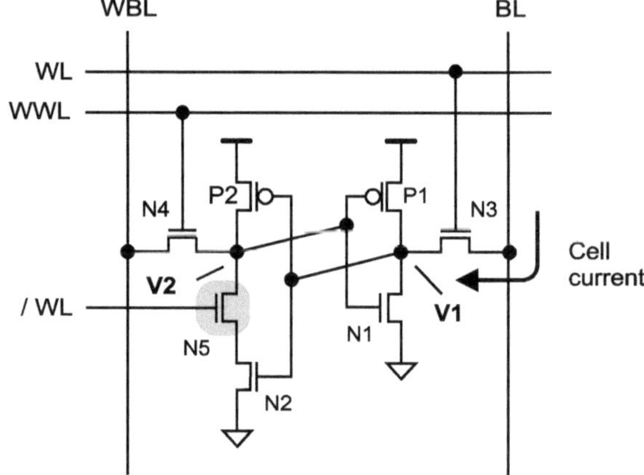

Fig. 3.15 7T bitcell with protection device to prevent read disturb [32]

to leakage current. In order to achieve stable read operation, the WL pulse width should be lower than the retention-time-to-fail due to leakage.

To push the SRAM operation down to subthreshold region, 10T bitcells have been proposed [33, 34]. Figure 3.16 shows a 10T subthreshold bitcell with fully differential read that can support column interleaving. In read mode, the WL is enabled and $VGND$ is pulled to ground while WWL is disabled. Therefore, the storage nodes Q and QB are isolated from the bitlines in read mode, and the read SNM is the same as the hold SNM of conventional 6T SRAM. In write operation, both WL and WWL are enabled to transfer the write data to the cell node. To improve write ability of 10T bitcells at low voltage, WWL boosting has been proposed.[1]

Recently, a 10T bitcell that allows contention-free writes and improves V_{min} has been proposed [34], as shown in Fig. 3.17. In this bitcell, complementary WL

[1] WL boosting is one of the write assist techniques that will be described in the next sections.

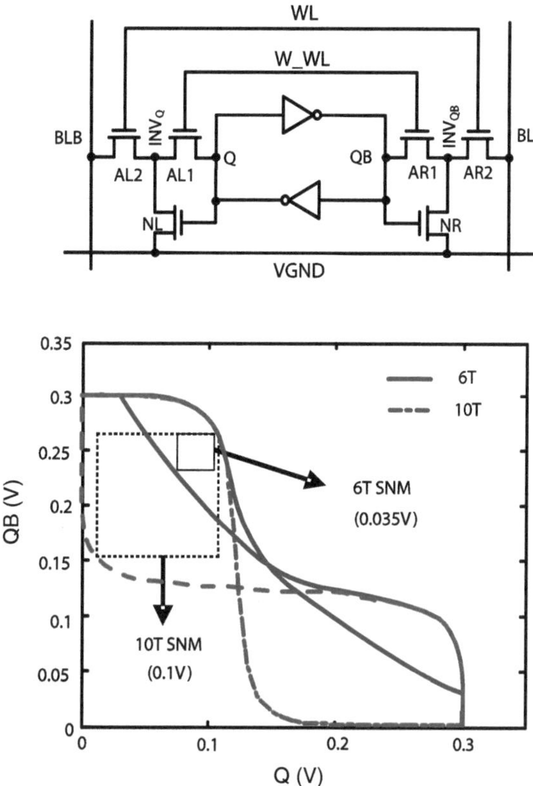

Fig. 3.16 10T bitcell that allows SRAM operation down to subthreshold region [33]

($WRWL$ and $\overline{WRWL}$) are used to access the bitcell in write operation using a transmission gate consisting of NMOS and PMOS transistors. Also, the write wordlines are used to disable the bitcell internal pull-up and pull-down devices, allowing the write operation to be contention-free when the $WRBL$ is used to write the bitcell. The read operation uses a separate read port, similar to 8T bitcells [34]. In addition to the bitcells mentioned above, other types of bitcells have been proposed to improve V_{min}, such as asymmetrical 6T bitcell [24].

3.4 Write and Read Assist Circuits

As discussed earlier, one of the biggest challenges in designing SRAM bitcells is to achieve balanced write and read margins. This is because of the conflicting requirements on the bitcell design, which are difficult to balance by conventional sizing and V_{th} adjustments [35]. To address this challenge, extensive research has been done in

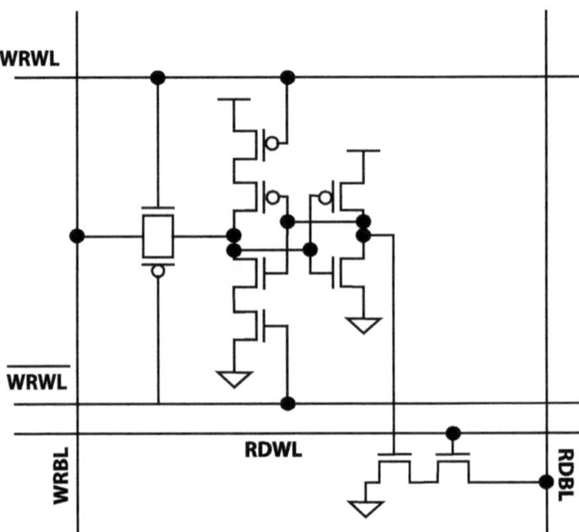

Fig. 3.17 10T bitcell that allows contention-free writes [34]

the area of write and read assist techniques to reduce the conflict between write and read operation and allow the memories to operate at lower supply voltage.

Read assist techniques include using higher supply voltage for the bitcell, wordline under-drive (WLUD), lower bitline capacitance, lower bitline voltage, modulating device characteristics using body bias, and read and write back [36, 37]. Since write operation in 6T bitcells has conflicting requirements with read operation, write assist techniques tend to perform the opposite operation compared to read assist [36, 37]. Write assist methods include lowering bitcell supply voltage, wordline boosting, negative bitline write, and body bias to improve the strength of NMOS pass-gate versus the PMOS pull-up. Assist techniques can be broadly classified into two categories: single supply and dual supply techniques. In the next sections, we discuss the implementation details of different state of the art write and read assist techniques [36–42].

3.5 Dual Supply Assist Techniques

In dual supply approaches, one power supply is used for the bitcell array, and a different supply voltage is used for the periphery circuit [30, 42–44]. In this way, the SRAM's supply voltage can be kept at a higher voltage compared to that of logic. The logic's supply voltage can be scaled to reduce power consumption, while the SRAM's supply is kept constant. This ensures that the SRAM read stability is sufficient, since SRAM failures can be eliminated at higher supply voltage (at higher V_{DD}, the impact of variations decreases). Many varieties of dual supply memories have been proposed [38, 39, 41, 42, 45–48].

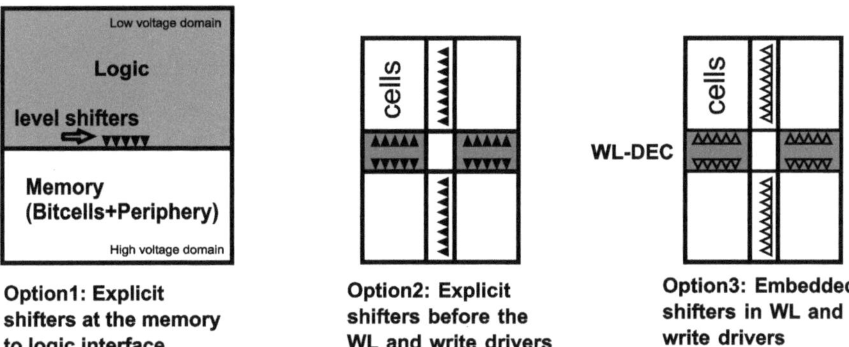

Fig. 3.18 Different options for partitioning the voltage domains in dual supply read and write assist techniques [30]

An important design decision related to dual supply assist is how to partition the voltage domains. Different types of partition have been proposed that cover different points in the tradeoff between memory area, power, and bitcell stability, as shown in Fig. 3.18. In the simplest approach, the whole memory (bitcell+periphery) lays on the high voltage domain, hence, level shifters are only needed in the interface between memory and logic [30]. The second option is to place the level shifters internal to the memory at the boundary between the bitcell and periphery, either using explicit level shifters or embedded level shifters, as shown in Fig. 3.20. The first option provides the smallest area since it involves the lowest number of shifters; however, the power consumption is the highest because the memory periphery is at the high voltage domain and cannot be scaled. Using level shifters inside the memory, the power consumption can be reduced by 20–30 %, and the use of embedded level shifter further reduces the area overhead [30]. Figures 3.19 and 3.20 shows the implementation of level shifters in the BL and the WL drivers, respectively [30]. These types of embedded level shifters help to reduce the memory area compared to using explicit level shifters, and prevent leakage paths in the boundary between the low and high voltage domains.

In concept, designing the memory so that the read and write operations use different supply voltages significantly improves cell stability. In read operation, WL voltage is lower than the array voltage, which increases SNM (pass-gate drive capability decreases). In write operation, the WL voltage is higher than the array voltage, which improves the bitcell write margin [48]. The challenge is that in high density memories, column interleaving is often used to improve area efficiency and reduce soft error rate. Hence, in both read and write operations, bitcells that are on the same wordline will experience a dummy read operation (half-selected bitcells). Therefore, the simple implementation of static dual supply along the wordline direction cannot address read and write stability for column interleaved memories.

To address the issue of read and write for column interleaved memories, column-based dual-power supply has been proposed [35, 45, 49]. The bitcell's dynamic dual

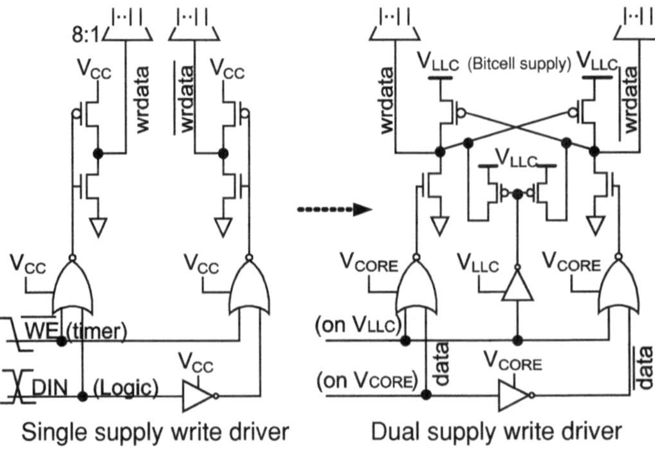

Fig. 3.19 Embedded BL level shifter implementation in dual supply memories [30]

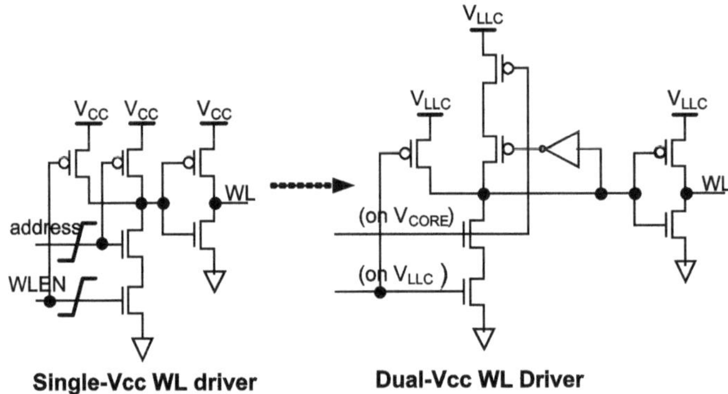

Fig. 3.20 Embedded WL level shifter implementation in dual supply memories [30]

supply is switched during read and write operations at a column level as shown in Fig. 3.21. To improve read stability, a higher VCC is connected to the memory cells in read operation, which increases the cell SNM. In write operation, a lower voltage is connected to the memory cell which makes the cell easier to flip. The change from high to low voltage is accomplished using a column-based power multiplexer (mux) as shown in Fig. 3.21. Therefore, only the column that is being written experiences low voltage V_{CC_lo}, while half-selected bitcells that are on the wordline stay connected to V_{CC_hi}, which ensures that the read stability for those cells is not reduced. The low supply voltage needs to be higher than the retention V_{min} to prevent bitcells on the column being written from losing the stored data. Another key point in this approach is that the local VCC needs to be stabilized before the WL is turned on to prevent any stability issues, which requires that the VCC select signals are generated early enough.

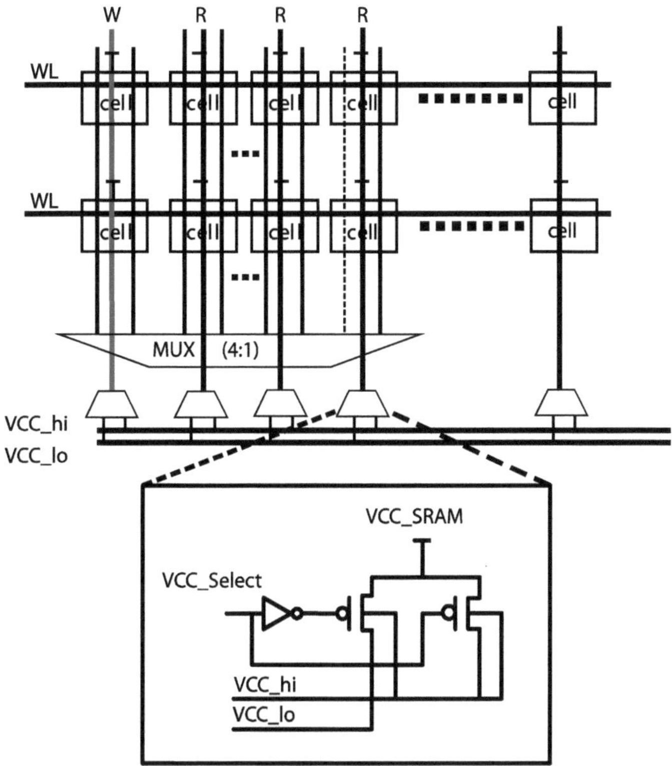

Fig. 3.21 Write assist using column-based dual supply [43]

In dual supply read assist techniques, the bitline can be precharged to the high [30, 39] or low voltage [42, 50]. Figure 3.22 shows the impact of lowering the bitline voltage on read SNM. As the bitline voltage decreases, the strength of the NMOS pass-gate decreases, hence, the SNM initially improves. However, as the bitline voltage decreases further, SNM starts decreasing after a certain point since the bitcell experiences a dummy write operation. This limits the minimum precharge voltage of the bitlines, and the power reduction achieved by lowering the logic voltage [50]. Figure 3.23 shows an implementation of dual supply read assist with the WL and bitcell supply at high voltage while the bitlines are precharged to the low supply voltage. One way to extend the range of the low supply voltage is to have the cell voltage track the logic voltage, while still maintaining higher voltage than the bitlines [50].

To improve the effectiveness of dual power supply SRAM, programmable WL voltage level control has been proposed [45]. In this technique, both the low and high supply voltages are provided to the WL drivers, which are level programmable using a digital code, as shown in Fig. 3.24. The WL voltage level is controlled adaptively depending on the process corner. So that the WL control balances between read disturb and write failure, as shown in Fig. 3.25

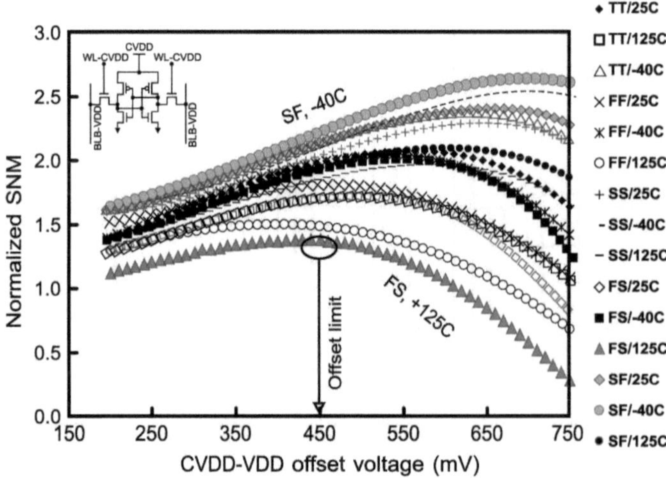

Fig. 3.22 Impact of lowering the bitline voltage on read SNM for different process corners and temperatures in 45 nm technology [50]

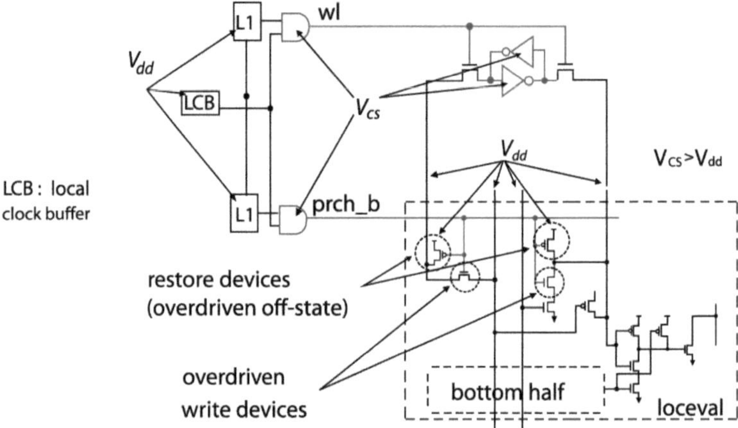

Fig. 3.23 Implementation of dual supply read assist with the WL and bitcell supply at high voltage while the bitlines are precharged to the low supply voltage [42]

Another technique that utilizes dual supplies use two voltage levels to control the body bias [44]. Using the body bias for the NMOS and PMOS separately can improves the SRAM margins, since SRAM margins are sensitive to the ratio of NMOS to PMOS drive. In column-based body bias, the PMOS pull-up is forward biased in read operation to improve read stability. In write operation, for the columns selected for write, the body bias is reversed, which weakens the pull-up device and improves write ability. Half-selected bitcells stay at the forward-bias conditions, which is the same as in the read operation.

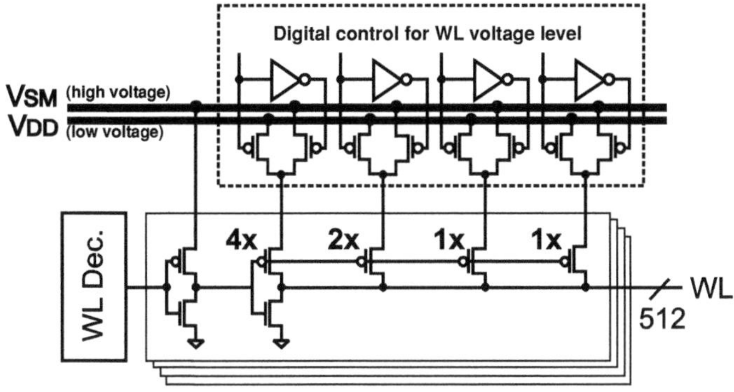

Fig. 3.24 Dual supply read assist using adaptive WL voltage level control [45]

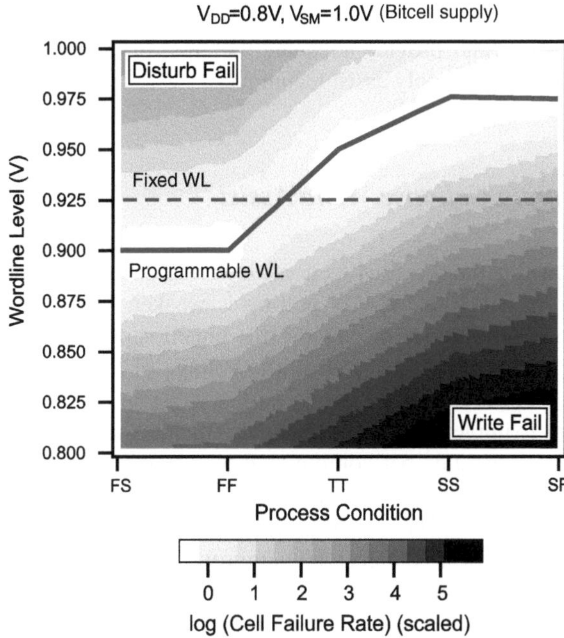

Fig. 3.25 Programmable WL used to track the process corner and apply the optimal WL voltage to balance between read and write failures [45]

3.6 Single Supply Write and Read Assist Techniques

Although the dual V_{DD} concept seems relatively simple, it introduces significant challenges. First, voltage level shifters are required at the interface between the bit-cell array voltage and the logic voltage. These level shifters tend to consume large area which lowers the memory's area efficiency. In addition, level shifters introduce additional delay in the memory critical path and hence, cause speed penalty. Moreover, the power grid design at the chip level becomes challenging since a dedicated power grid is required for memories, which adds cost due to routing and extra metal resources required to distribute the additional power supply.

The degree of difficulty in implementing a dual supply power grid depends on the chip architecture. For example, microprocessor designs use relatively few kinds of SRAM architectures having large capacity, such as caches, which can be physically placed in a close proximity on a chip [22]. This simplifies the design of a dual power grid since all the memories are physically located near each other. However, in SoC design, there are typically hundreds of SRAM architectures and they are not necessarily placed in close proximity on a chip. This makes it difficult to have a dedicated power grid for all the memories. Therefore, for an SoC, it is always desirable to use a single power supply for the SRAM [5, 22, 51, 52].

In single supply assist techniques, additional circuitry is added to assist write and read operations and provide adequate margins. In the next section we present an overview of single supply assist techniques.

3.6.1 Supply Collapse Write Assist

As mentioned in the previous sections, lowering the bitcell supply voltage improves write margin, since the PMOS pull-up is weaker. Several implementations have been proposed to reduce or collapse the bitcell supply voltage in write operation. These techniques rely on lowering the voltage of the bitcell supply during write operation [45, 53–56]

One way to implement supply collapse is to float the bitcell supply power [53]. The V_{dd} lines are separated per column ($Vddm[n]$) as shown in Fig. 3.26. $Vddm[n]$ is controlled using the power switch $MSW[n]$, which is turned off in write operation, so that $Vddm[n]$ floats. The bitcell write current (from $Vddm$ to the bitline pulled low) discharges the $Vddm$ voltage, which improves the write margin [53]. The reduction of $Vddm$ voltage may cause the bitcells on the same column to lose data due to retention failures. Therefore, careful design of this technique is needed to prevent bitcells from failing due to retention. In addition to floating bitcell supply, actively pulling down the supply has been proposed [55, 57, 58] as well as using charge sharing to reduce the supply voltage [51].

Recently, the concept of collapsing the supply voltage had been extended to the bitcell supply voltage below the static data retention voltage [55, 59]. To prevent the resulting data retention failure of unselected cells, the duration of voltage collapse must be controlled, requiring a dynamic data retention approach. The maximum time within which the data is retained in the bitcell after fully collapsing the supply voltage Td, max has been measured as shown in Fig. 3.27. The measured data shows that bitcells can retain data even if the supply voltage is fully collapsed as long as the collapse time is less 3 ns and 10 ns for the SP and LP cases, respectively.

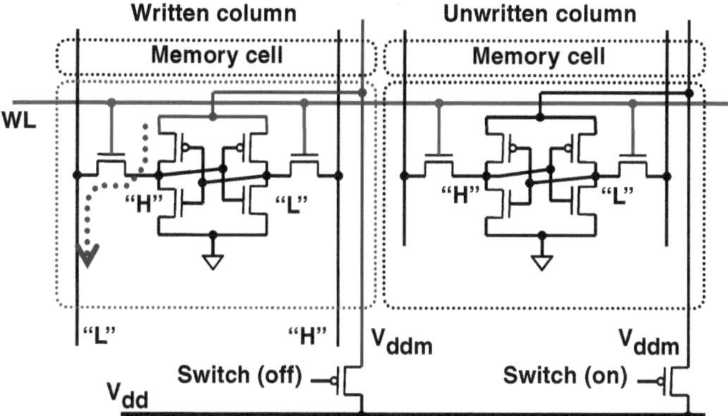

Fig. 3.26 Single supply write assist by floating the bitcell supply in write operation [53]

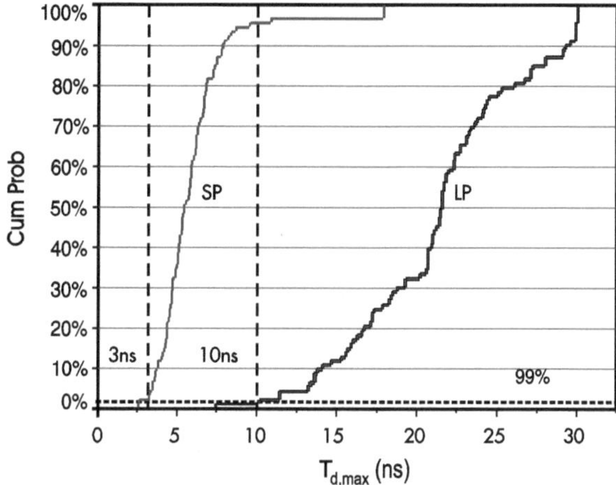

Fig. 3.27 Distribution of the maximum time a bitcell can sustain a collapsed voltage without flipping ($T_{d,\max}$) in 32 nm. SP and LP are two process flavors; LP being the low power option. Measured $T_{d,\max}$ show that bitcells can sustain 3 ns and 10 ns of supply collapse at the 99th percentile for SP and LP options, respectively [59]

To take advantage of dynamic data retention, transient voltage collapse write assist (TVC) has been proposed [55]. In this approach, the write assist lowers the bitcell supply below the retention voltage during write operation, hence, eliminating the contention between the pass-gate and the pull-up. Figure 3.28 shows the implementation of the TVC technique. In write operation, the header device ($PVCS$) is disabled and the cell supply (VCS) is actively discharged via the NMOS pull-down (NWR). The slew rate and minimum voltage of VCS can be controlled by varying the pulse width $COLPULSE$ and the parallel PMOS clamp devices PB. This technique requires careful control of timing and the lowest VCS voltage to ensure that the collapse time is sufficient for write without causing retention failures on unselected cells [55].

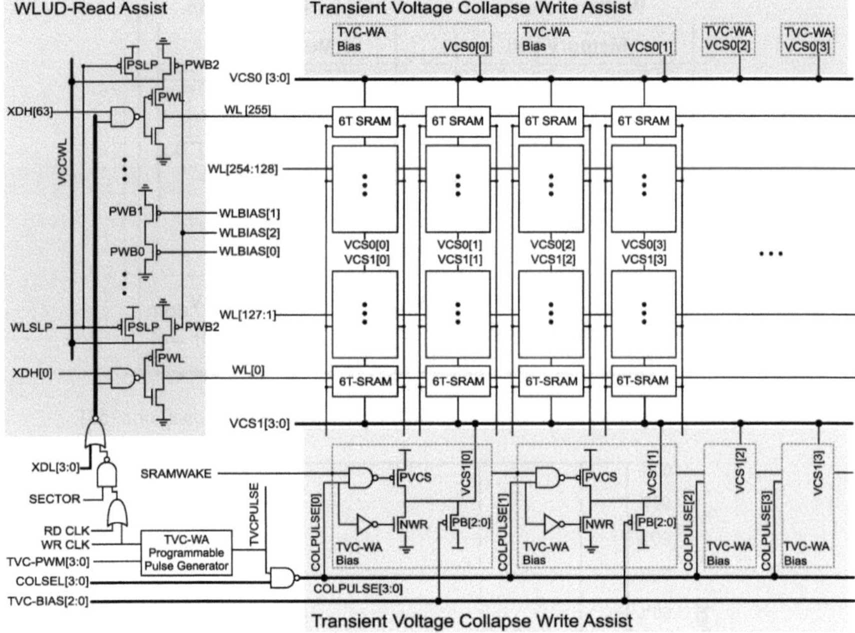

Fig. 3.28 Transient voltage collapse write assist and wordline under drive read assist [55]

Other supply collapse techniques include charge sharing to reduce the supply voltage. Also self-induced supply collapse was proposed to improve the write margin for 8T bitcell [54]. The scheme pre-discharges the bitlines prior to write operation: the virtual bitcell supply slightly droops before the start of the write operation. A similar idea to lowering the supply is to boost the virtual ground [2], but it requires larger series devices in the Vss path due to higher current demands [55].

3.6.2 *Negative Bitline Write Assist*

Another write assist technique relies on improving the strength of the NMOS pass-gate by applying a small negative voltage [60–64]. Using a small negative bitline voltage (~200 mV) the gate-source bias of the pass-gate increases, which increases the write margin, as shown Fig. 3.29. The negative voltage is typically generated using a boosting capacitor connected to the bitlines and enabled at the end of the conventional write operation (after the bitline is pulled down to ground). Large improvements in write margin, in the range of 2–3 orders of magnitude reduction in failure probability can be achieved using this technique as shown in Fig. 3.29.

Figure 3.30 shows one implementation of the negative bitline technique [63]. In this approach, after the bitline is pulled low, the bitline is floated by disabling

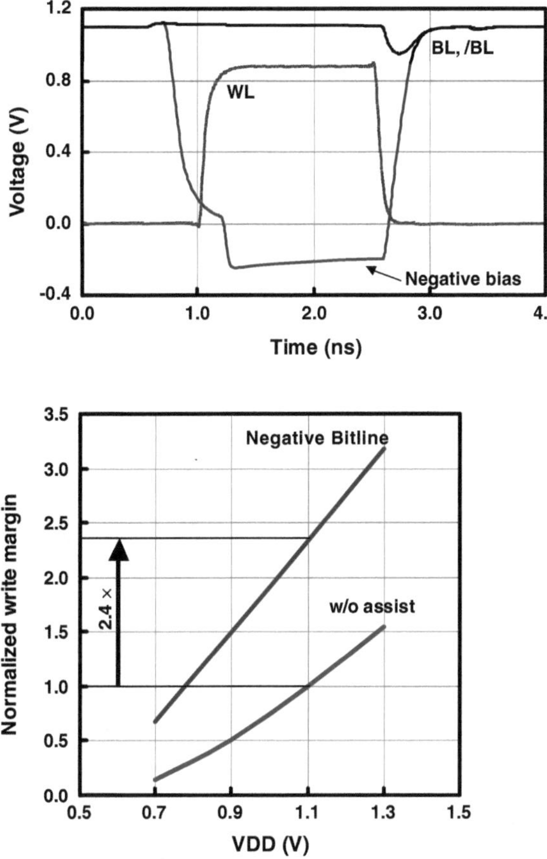

Fig. 3.29 Concept of negative bitline to improve write margin [60]

the column select devices. By using a high to low transition on the other side of the coupling capacitor C_{boost}, the bitlines are temporarily transitioned to negative bitline. The ratio of C_{boost} to total bitline capacitance determines the coupling coefficient and the magnitude of the negative voltage. The value of the negative bias should not be excessive to prevent flipping bitcells on the same column. This may occur because the pass-gate of non-selected bitcells may turn on if the bitline negative voltage exceeds the threshold voltage of the pass-gate, causing a retention failure. To allow memory configurations that have different number for cells per bitline, improvements in this technique use automatic BL negative level adjustment [61]. These improvements limit the value of the negative voltage, allowing this technique to be suitable for compilable memories.

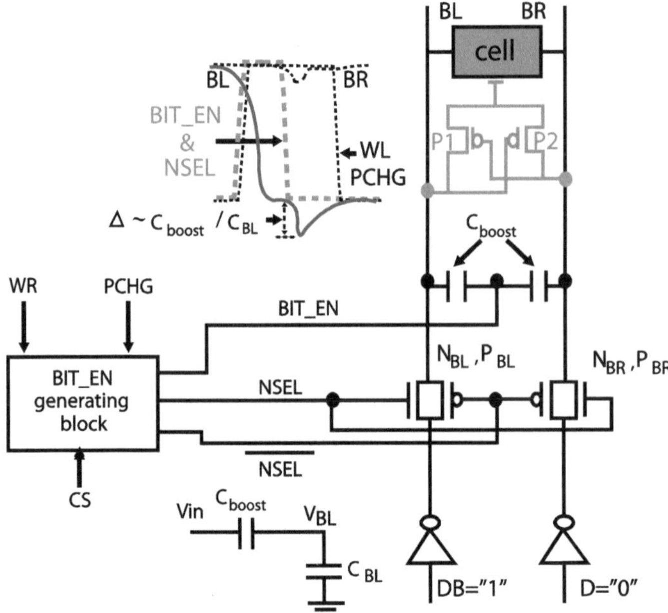

Fig. 3.30 Circuit implementation of negative bitline write assist [63]

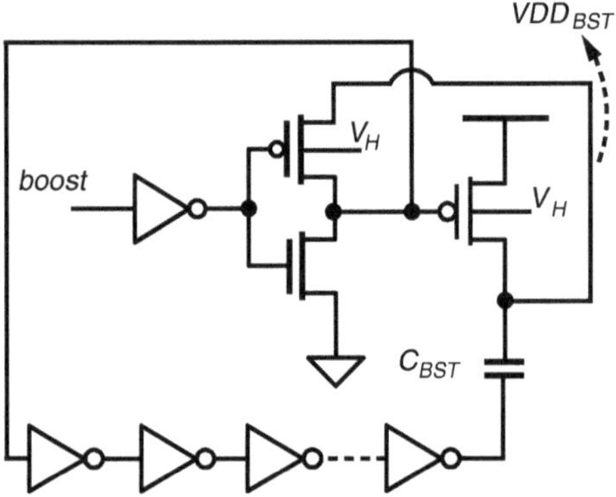

Fig. 3.31 Write assist using WL boost to improve write ability in 8T bitcells [26]

3.6.3 Wordline Boosting Write Assist

Wordline boosting technique is another approach to improve bitcell write ability. This technique is typically used with 8T bitcells that do not support column interleaving, and therefore, do not suffer from the half-select problem [26]. Boosting WL effectively reduces the write V_{min} for the 8T bitcells. One way to boost the write

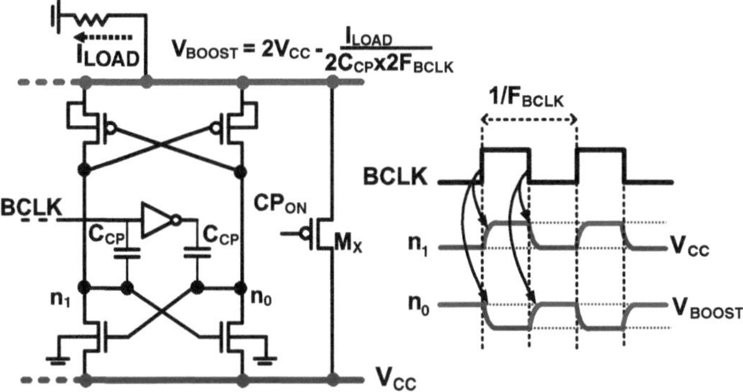

Fig. 3.32 Wordline boosting write assist technique employing a charge pump to boost wordline voltage in write operation [52]

wordline WWL is to use a boosting capacitor, as shown in Fig. 3.31. Boosted voltage ($V_{DD,BST}$) powers the WL drivers, the write drivers, and the column select devices during write operation [26].

Another approach to boost the WL is to use a charge pump [52] as shown in Fig. 3.31. The pump boosting ratio is determined by the load current, which comprises all the active and inactive level-shifters connected to the WL. In this implementation, if the boosted voltage exceeds the V_{max} of the device, the charge pump is turned off to maintain the gate oxide and junction reliability of the devices connected to the V_{boost} rail. Since the boosting ratio depends on the pump frequency, it is important to track the process and temperature condition to provide the optimal boost. On-chip write ability sensors that dynamically track the temperature and aging has been used to provide the optimal boosting ratio [52]. Boosting techniques using charge pumps consume large power consumption due to the dynamic power of the charge pump and the need to increase the pump frequency (F_{BLK}) to improve the boosting ratio [52] (Fig. 3.32).

Recently, a novel WL boosting approach has been proposed which uses the intrinsic coupling capacitance from the device and interconnect to the WWL to create a boosting capacitor, as shown in Fig. 3.33. The technique uses two types of intrinsic coupling capacitance. The first coupling capacitance is found in the WWL interface of the WWL driver ($C1$), while the second coupling capacitance is at the WWL interface to the bitcell NMOS write pass-gate ($C2$ and $C3$). To enable the use of the coupling capacitance in the WWL driver, the input of the driver is transitioned to low to create the rising transition on the WWL. After the WWL reaches full V_{DD}, the WWL is floated, and the coupling capacitance is enabled by switching the $BOOST$ signal from low to high. Similarly, the second coupling is enabled by pre-discharging the WBL and $WBLX$ signals, and depending on the data polarity, one bitline is brought high after the WWL has been floated. This technique can achieve a total boost of 20 % of the supply voltage, with the pass-gate coupling contributing 17 % while the WWL driver contributes 3 %. This boosting technique is scalable

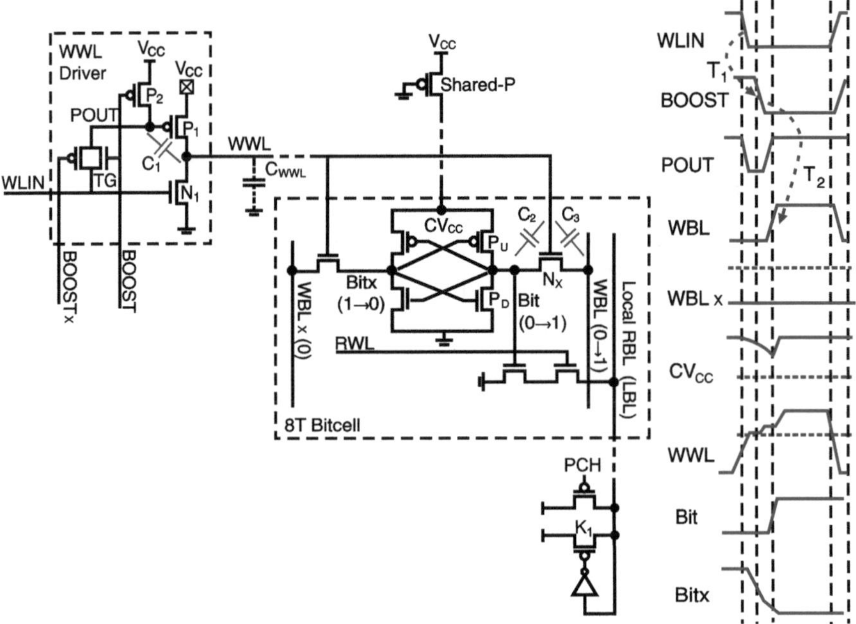

Fig. 3.33 Write assist employing wordline boosting using intrinsic capacitive coupling from (1) the bitlines to the WL and (2) WL driver input to the WL output [54]

to any number of bits per WWL since the intrinsic bitline capacitance scales in the same ratio.

3.6.4 Wordline Under-Drive Read Assist

Lowering the wordline voltage in read operation decreases the pass-gate strength and improves the read stability. This technique is one of the most commonly used read assist techniques [22, 45, 51, 58, 65, 66]. Figure 3.28 shows an implementation of WLUD, where the WL voltage is adjusted using a voltage divider between the PMOS PWL device and the programmable pull-down $PWB1, 2, 3$.

The wordline under drive technique is enabled in both read and write operation so that half-selected bitcells experience lower WL voltage. However, lowering the WL voltage degrades the write margin, so a write assist technique is usually implemented with the WLUD approach. In addition, the WL level should be controlled depending on the process and temperature condition, so that the lowest WL voltage occurs in the worst global corner for read stability (fast NMOS and slow PMOS), while the WL voltage stays high in other corners which are write-limited. To address this requirement, adaptive control of the wordline level depending on the process corner has been proposed [22, 45]. In [22], WL voltage varies depending on D2D and temperature variations. Replica transistors are used to control the WL voltage level, which improves process tracking and increases SNM. A bitcell-based sensor has been proposed to dynamically optimize the level of the WL for each die [65]. The on-die sensor determines if the die is read or write-limited, and a programmable control applies the optimal WLUD. Since read and write stability strongly depend on the temperature, a die can shift from being write-limited at low temperature to read-limited at higher temperature. The sensor dynamically tracks the temperature and process variation and applies the optimal WLUD level [65].

3.6.5 Lower Bitline Read Assist

As discussed in Sect. 3.5, lowering the bitline voltage level improves read stability since the pass-gate strength decreases [2, 11, 29, 64, 67]. In one such approach, the bitline precharge voltage is reduced before the bitcell is accessed, increasing SNM. The bitline voltage can be reduced by slightly discharging the bitlines before the start of read operation [29]. In another implementation, the bitlines are precharged using an NMOS transistor, so the precharge voltage is one V_{th} lower than V_{DD}. Another technique precharges different columns to V_{DD} and GND and uses charge sharing to reduce the bitline voltage [67]. More details about this approach will be discussed in the case study presented in Sect. 3.7.

3.6.6 Short Bitline Read Assist

Dynamic read stability analysis revealed that reducing the bitline capacitance reduces read disturb failures. Lowering the number of cells per bitline from 256 to 64 cells per bitline reduces failure[2] rate by 50X, as shown in Fig. 3.34. Reducing the number of bitcells per columns is therefore an effective approach to improve read stability and has been explored to reduce SRAM V_{min} [17, 26].

3.6.7 Read and Write Back Assist

Several read assist techniques that rely on improving the dynamic stability of the bitcell have been proposed. The duration of the WL pulse width has a strong impact on read and write stability. If the WL pulse width is very short, the dynamic read

---

[2] The DC failure rate is 500X higher than the rate compared to dynamic failure rate for the 64 cells per bitline case. Also, DC failure is 140 mV higher than dynamic failure, which shows the importance of dynamic read stability.

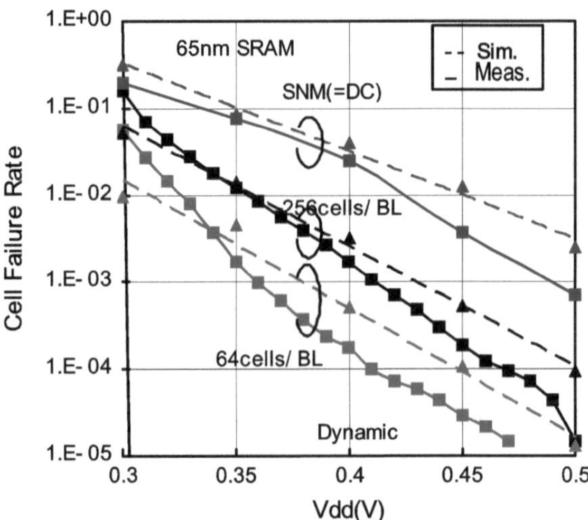

Fig. 3.34 Measured static and dynamic read stability versus supply voltage. Reducing the number of bitcells per columns reduces the probability of read disturb [24]

stability improves since the bitcell is not given enough time to flip. However, the write margin worsens for the same reason [29]. Therefore, an optimal WL pulse width balances read and write stability. To decouple the opposite requirements of WL pulse width for read and write operations, the read and write back approach has been proposed (also known as read modify write) [29, 49]. In this technique, each column has a dedicated sense amplifier (column multiplexer=1). As the name implies, the write operation starts with a read operation followed by a write operation. In the read operation, a narrow WL pulse width is used, to increase read stability. The read data is stored on a per column sense amplifier, which acts as the write driver. Next, the WL is enabled again for a longer pulse width that is sufficient for write operation. The same concept can be applied in read operation only, where the sense amplifier is used to read the data from the bitcell and write it back at the end of the read operation [49]. Hence, it provides data recovery by writing back the original data. Read and write back technique increases power consumption since every column undergoes full signal amplification. Moreover, it has a large area overhead (8 % area increase) since a SA is integrated per column and cannot be shared between several columns as typically is the case with high density memories [29]. In addition, mismatch in the sense amplifier may cause the bitcell to write incorrect data, corrupting the stored one.

3.7 Case Study: Selective Precharge Read Assist Technique

In SRAM design, the bitlines are typically precharged to V_{DD} before accessing the bitcell. However, as discussed earlier, reducing the bitline voltage V_{BL} before accessing a bitcell improves read stability, as shown in Fig. 3.35. This increase in

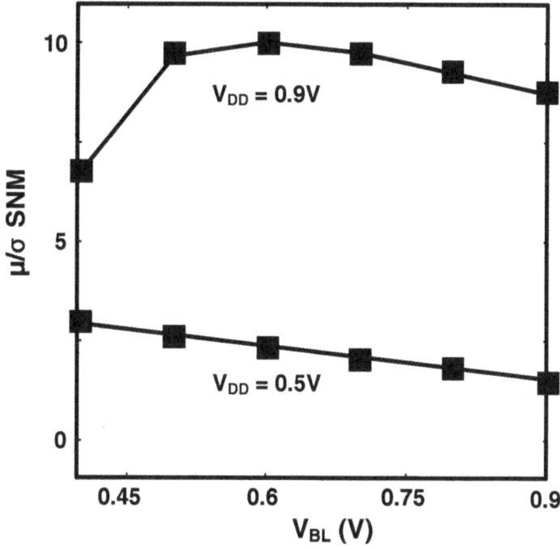

Fig. 3.35 $\frac{\mu}{\sigma}$ SNM versus bitline precharge voltage. Lower bitline voltage during a read access improves bitcell read stability (SNM) [11]

stability results from the decrease in pass-gate (access transistor) strength reduces as the bitline voltage decreases. This effectively increases the bitcell α ratio, defined as the ratio of pull-down to pass-gate strength. Hence, SNM improves as ΔV_{BL} increases where ΔV_{BL} is defined here as $V_{\mathrm{DD}} - V_{\mathrm{BL}}$ just before the wordline (WL) is asserted. Note that as ΔV_{BL} increases,[3] SNM reaches a maximum, and further increases in ΔV_{BL} cause significant SNM reduction. Therefore, accurate control of ΔV_{BL} is important to prevent ΔV_{BL} from exceeding the maximum SNM point.

From a circuit point of view, the relation between ΔV_{BL} and SNM has been exploited into increase bitcell stability [2, 29] . A pulsed bitline approach has been used to control the duration of an NMOS pull-down [29]. This pull-down device discharges the bitline, which increases ΔV_{BL} just before the WL is enabled. However, this technique is sensitive to PVT variations since ΔV_{BL} is a strong function of the pulse duration, which will vary with PVT variations. Therefore, a complex timing scheme may be required to control ΔV_{BL} at different PVT conditions. Alternatively, an NMOS device has been used to precharge the bitlines, by having one V_{th} drop [2]. Due to the strong sensitivity of V_{th} to PVT variations, the effectiveness of this technique decreases in different PVT corners. Moreover, low V_{th} devices are required to ensure that ΔV_{BL} does not cause read disturbs in the worst-case conditions, which adds additional processing cost, especially for low cost SoCs.

In this section, we examine implementation details of a read assist technique that uses lower BL voltage to improve read stability [67]. This technique uses a single supply voltage approach to overcome the obstacles in implementing dual supply read assist techniques. Instead of precharging the bitlines to V_{DD}, different parts of the bitlines are precharged to V_{DD} or (predischarged) to GND. Using charge sharing, the final required bitline voltage can be precisely controlled using the capacitance ratio. Therefore, this technique is highly immune against process variations (both

[3] ΔV_{BL} in this chapter should not be confused with the bitline differential voltage ΔV_{bl}. Here, ΔV_{BL} is the reduction in bitline precharge level before accessing the bitcell.

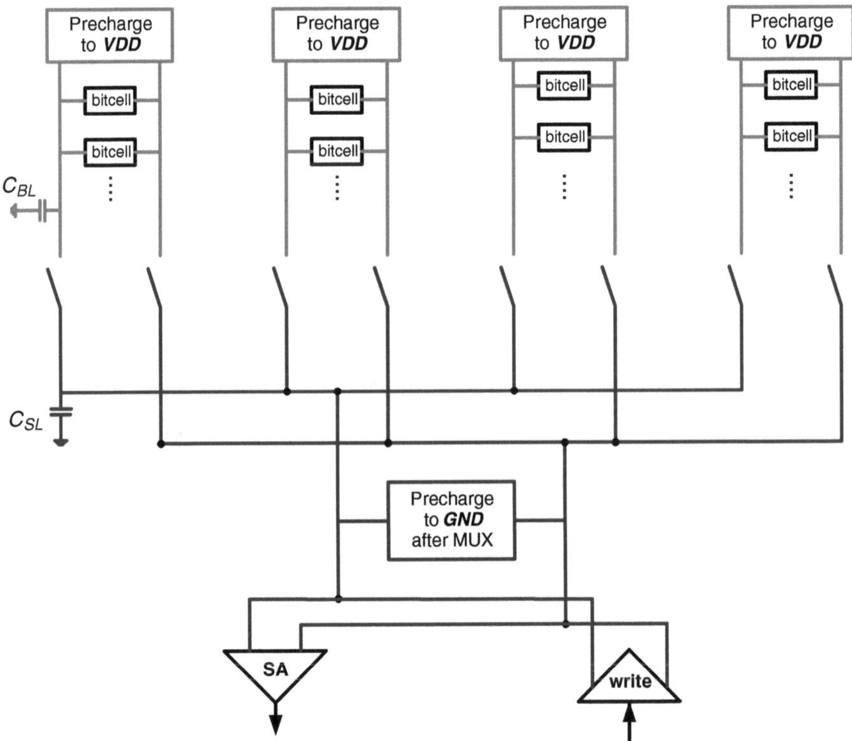

Fig. 3.36 Selective Precharge operation. Step 1: Precharge to V_{DD} and GND

front-end and back-end) since the capacitance ratio is weakly dependent on PVT corners.

3.7.1 Circuit Operation

Figure 3.36 shows a simple schematic for the selective precharge technique with four bitline columns connected to the read and write circuitry (sense amplifier and write drivers). Bitlines, BL/BLB, refer to the upper part of the bitlines connected directly to the bitcells (before the column select). Sense/Write lines, SL/SLB, refer to the lower part of the bitline connected to the sense amplifier and write drivers (after the column select).

Selective precharge operation can be divided into three main steps. First, BL and BLB are precharged to V_{DD} as in conventional approaches, while SL and SLB are pre-discharged to GND, as shown in Fig. 3.36. In the second step, the column select devices (MUX) on each bitline column are enabled as shown in Fig. 3.37. Hence, charge sharing occurs between the upper and lower bitlines for $BL0 - BL3$. The

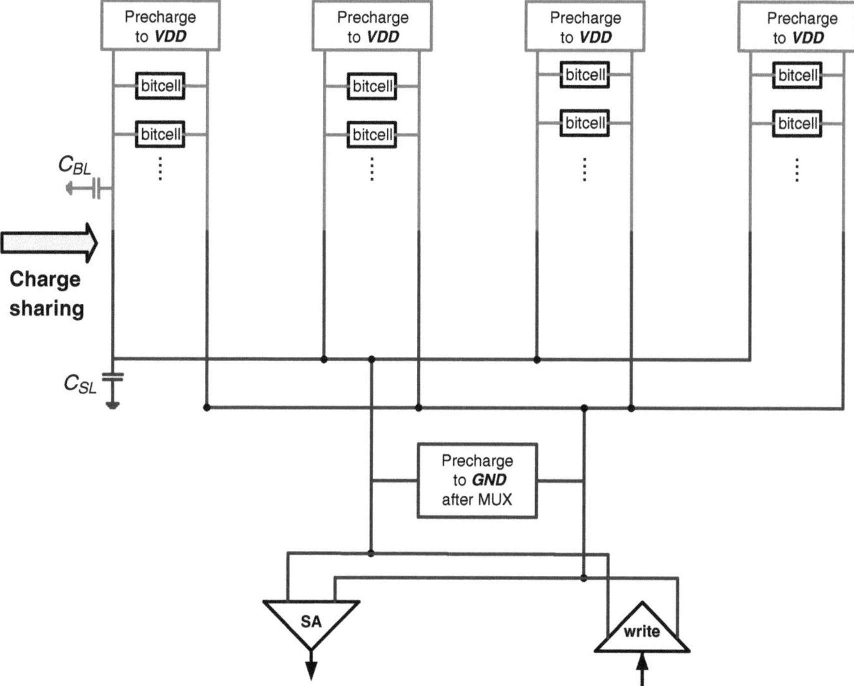

Fig. 3.37 Selective Precharge operation. Step 2: Charge sharing

final bitline voltage after charge sharing is determined by the capacitance ratio of upper to lower bitlines (BL and SL). Charge sharing reduces the bitline voltage so, SNM improves as discussed earlier. In the third step, the MUX devices for all unselected columns are disabled, while the selected column MUX stays on as shown in Fig. 3.38. In this case, the selected column provides access to the required bitcell, while SNM of half-selected bitcells is also improved since their bitline voltages have also been reduced.

Figure 3.39 shows the implementation of the selective precharge technique. A NOR gate is added for each bitline column to control the column select. Figure 3.40 shows the precharge circuits for both V_{DD} and GND. Figure 3.41 shows the timing diagram for selective precharge operation. *ch_sh* is activated using the rising edge of the precharge disable (for PMOS pull-up). When *ch_sh* is high, the PMOS devices in the column select MUX are on. Hence, charge sharing occurs between all bitlines sharing the same read/write circuitry and SL/SLB line. Therefore, BL/BLB voltage decreases while SL/SLB voltage increases. *ch_sh* is disabled using *mux_state*, which is a dummy column select signal. Therefore, bitcells see reduced bitline voltage when the bitcell is accessed. At the end of operation, BL/BLB are precharged back to V_{DD} while SL/SLB are precharged to GND, as shown in Fig. 3.41.

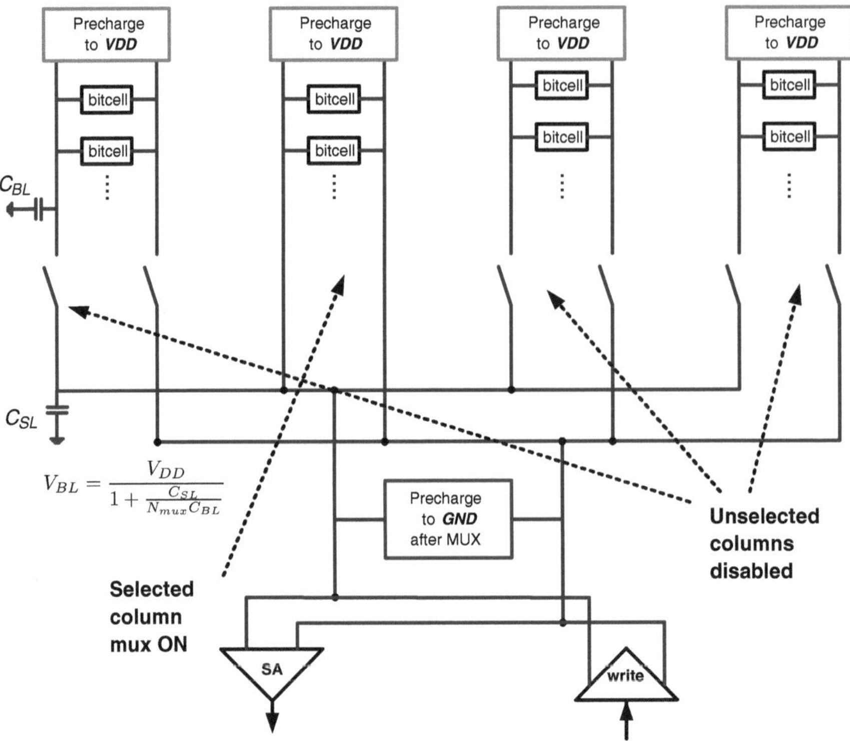

Fig. 3.38 Selective Precharge operation. Step 3: Unselected columns disabled

Selecting the location of precharge to V_{DD} or GND sets the required value of ΔV_{BL}. For example, if a larger ΔV_{BL} is required, one or more of the bitlines can be precharged to GND instead of V_{DD}, as shown in Fig. 3.42. Therefore, the proposed technique allows changing ΔV_{BL} by selecting in which points should be precharged to V_{DD} or GND. Note that in this technique, no additional supply voltages are required to generate the desired bitline voltage, which reduces the design complexity. In addition, since the final ΔV_{BL} voltage depends solely on capacitance ratio, its value is not influenced by process variations.

3.7.2 Access Time Improvement

In SRAM, the read operation determines the access time of the memory. One of the limiting factors determining the access time is the delay from the clock to the WL enable, which is part of the memory critical path. The proposed technique introduces another signal, ch_sh, which should be enabled before WL is asserted. To accommodate the ch_sh signal shown in Fig. 3.41, the WL enable path may

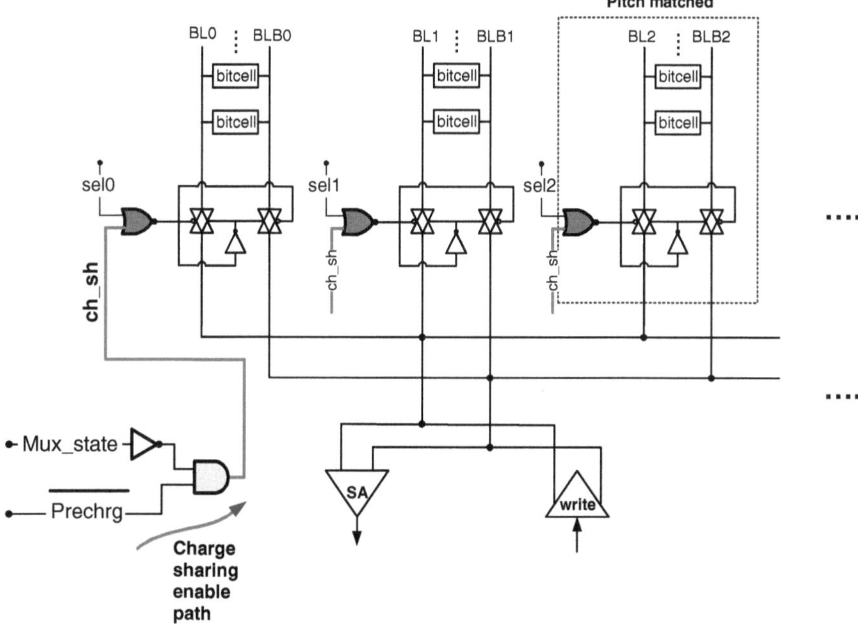

Fig. 3.39 Selective precharge schematic

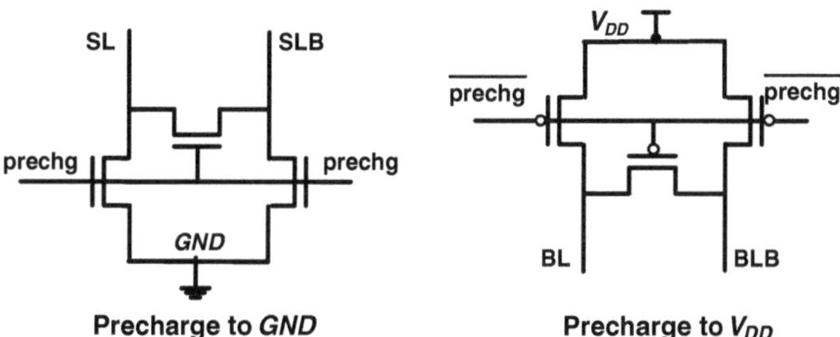

Fig. 3.40 Precharge to V_{DD} and GND circuits, including equalize transistors

be delayed. This delay will therefore increase the memory access time. Hence, a technique to reduce (or recover) access time is required.

In addition to the clock to WL delay, another contributor to the memory's access time is the WL pulse width T_{WL}. T_{WL} is the time required for the bitcell to discharge the bitlines and generate sufficient input differential for the sense amplifier to allow correct read operation. T_{WL} typically contributes to approximately 30 % of the memory access time [44]. To reduce this delay component, we exploit the relation between ΔV_{BL} and the sense amplifier (SA) input offset.

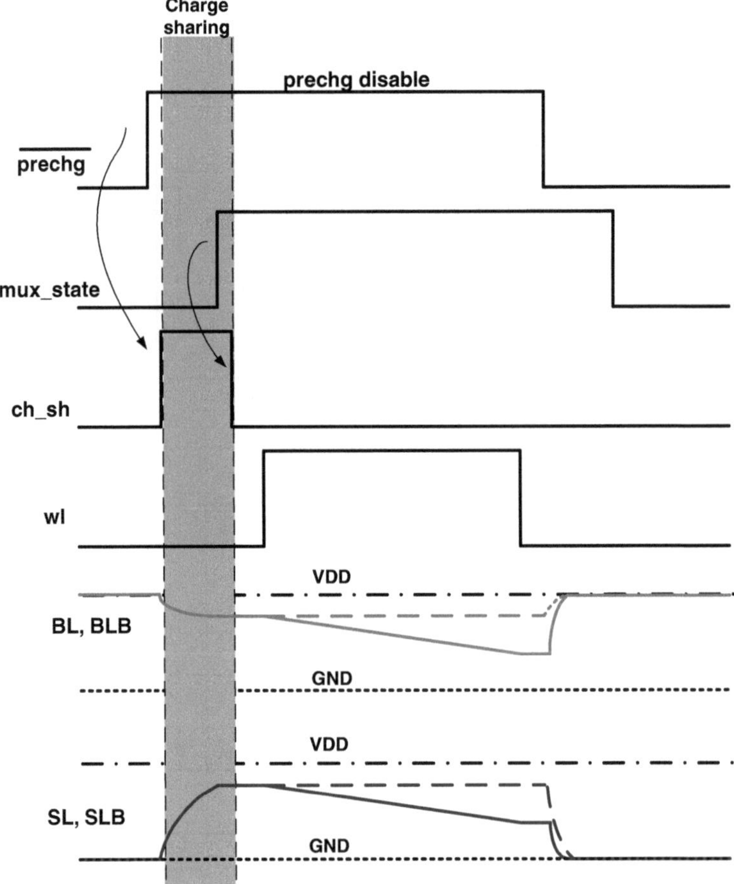

Fig. 3.41 Selective precharge timing diagram

There are many types of sense amplifiers used in SRAM design. However, current latch sense amplifier (CLSA) is one of the most widely used types due to its high speed and isolation as discussed in Sect. 5.4.2 (CLSA shown in Fig. 5.11). Moreover, it has been shown that reducing bitline voltage (common mode) improves the SA offset [68]. This characteristic of CLSA makes it very attractive in the proposed selective precharge technique. By reducing the bitline voltage (increasing ΔV_{BL}), the SA offset ($\sigma_{\mathrm{SA,offset}}$) decreases, allowing T_{WL} to be reduced for a given failure probability. The reduction in T_{WL} can therefore compensate for the increase in clock to WL delay, as will be shown in Sect. 3.8.

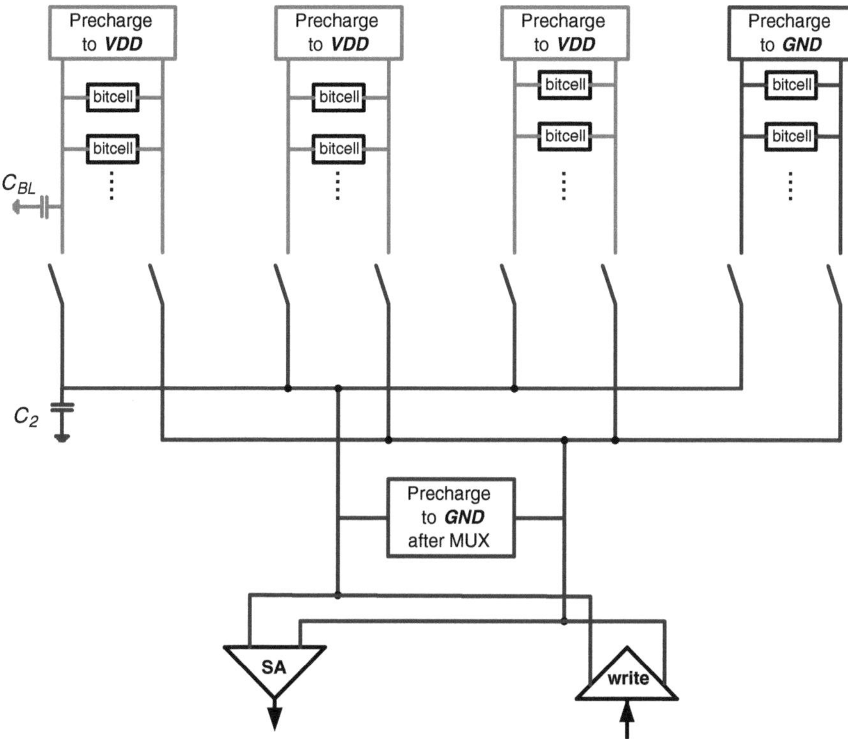

Fig. 3.42 Achieving larger range of ΔV_{BL} by precharging one of the bitlines to GND

3.8 Results and Discussion

To test the proposed read assist technique, a full-custom 512 kb SRAM was designed and implemented in an industrial 45 nm technology as shown in Fig 3.43. Table 3.2 provides details on the memory architecture. In this section, we present the post-layout simulation results for the proposed technique.

In Fig. 3.44, the read operation is shown for a bitcell on the first column (enabled using MUX0). In the beginning of the operation, $BL0$ and $BLB0$ are set to V_{DD} while SL and SLB are set to zero. Charge sharing operation is activated using ch_sh, which activates all the MUX transistors. Therefore, the $BL0/BLB0$ voltages decrease while SL/SLB increase as shown in Fig. 3.44, and they settle to a value determined by the capacitance ratio. Note that charge sharing happens quickly and that the results voltage is not sensitive to the ch_sh pulse width (a wider pulse does not affect the settling voltage after charge sharing). After charge sharing is completed, the MUX devices (PMOS) for all unselected columns are disabled (MUX1), while the selected column stays selected (MUX0). Therefore, when WL is asserted, the accessed and half-selected bitcells see a reduced bitline voltage, which increases the bitcell's SNM.

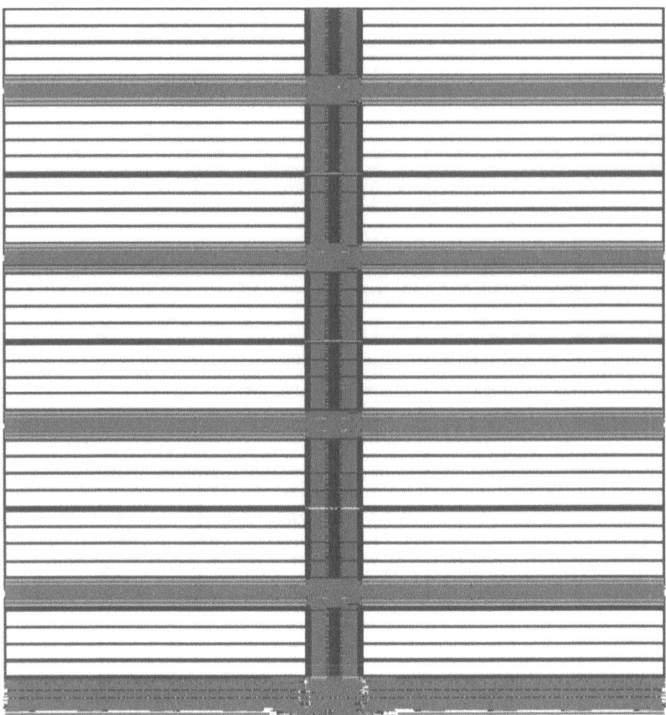

Fig. 3.43 Layout of the designed 512 kb memory in 45 nm technology

At the end of the read operation, the bitlines are precharged to V_{DD} while the sense lines are precharged to GND.

The impact of bitline voltage on the CLSA speed and input offset is shown in Fig. 3.45. Monte Carlo transient simulations were used to measure the SA's offset. As ΔV_{BL} increases, the SA delay slightly decreases until it reaches a minimum. Beyond that point, the SA delay increases. In the meantime, the SA input offset (σ_{offset}) decreases monotonically with the increase in ΔV_{BL}. The reduction in σ_{offset} improves the robustness of the SA and decreases the probability of read access failures. Therefore, the WL pulse width can be reduced accordingly based on the following:

$$\frac{T_{WL2}}{T_{WL1}} = \frac{\sigma_{SA,offset2}}{\sigma_{SA,offset1}}, \tag{3.1}$$

where T_{WL} is the time allowed for the bitcell to generate the bitline differential before enabling the SA. This large reduction in σ_{offset} reduces the access time of the memory. Because T_{WL} is about 30 % of memory access time, and as shown in Fig. 3.45, SA offset can be reduced by up to 25 %, and access time improves by 7 %. In reality, to accommodate the ch_sh pulse, the WL enabled path may be slightly delayed, so this

Table 3.2 512 kb memory design information

Technology	45 nm low power (LP) CMOS
Density	512 kb
Memory width (word size)	64 bits
Memory depth (number of words)	8,192 words
Banks	16 (32 kb each)
Rows/bank	128
Columns/bank	256

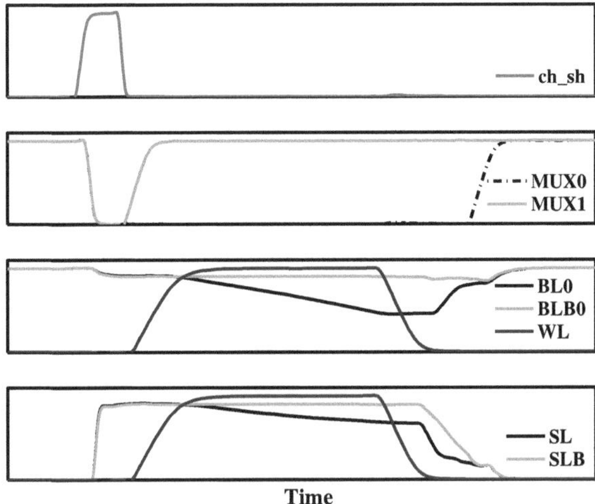

Fig. 3.44 Simulation waveforms for selective precharge read operation. MUX0/1 is the gate voltage for the PMOS device in the column select for column 0 and 1, respectively

improvement in speed would be smaller. Nevertheless, since charge sharing requires a very short time, the impact on access time improvement is negligible.

Charge sharing operation enabled when a bitcell is accessed for write operation, as shown in Fig. 3.46. In that case, half-selected bitcells experience reduced BL voltage to improve read stability (bitlines $BL1$ and $BLB1$). However, write margin degrades due to the lower drive capability of the pass-gate which may cause a write failure [39]. To improve the write ability of the selected bitcell, we use a CMOS write driver, as shown in Fig. 3.47. Therefore, the BL voltage lost in charge sharing is recovered using the pull-up device in the write driver. Hence, the CMOS write driver protects the write margin in write operation.

To estimate the bitcell read stability, we simulated the butterfly curve as shown in Fig. 3.48, which shows the bitcell SNM for a nominal bitcell which does not include WID variations. However, as discussed in Chap. 2, WID variations will cause

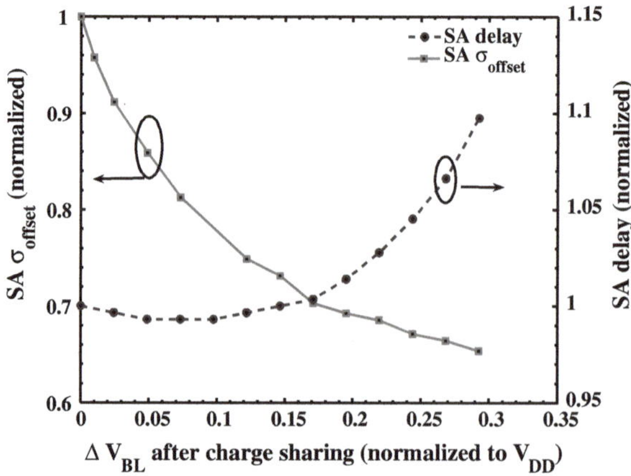

Fig. 3.45 Sense amplifier delay and input offset versus ΔV_{BL} after charge sharing

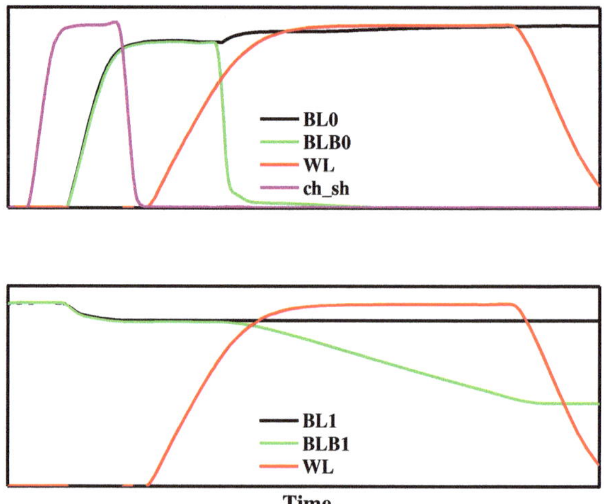

Fig. 3.46 Simulation waveforms for selective precharge write operation. $BL0/BLB0$ are accessed for write operation while $BL1/BLB1$ are half-selected bitlines

each transistor in the bitcell to have different V_{th}, which will cause the bitcell be asymmetric. This is shown by the Monte Carlo simulation results in Fig. 3.49, which shows a large spread in the VTC characteristics of the bitcell. This spread translates to large variation in SNM.

The improvement in SNM using the proposed technique is shown in Fig. 3.50. Monte Carlo simulations are used to measure the impact of local variation on the bitcell's SNM. To ensure high yield for the embedded memories, 6σ of SNM local variation is included. As ΔV_{BL} increases, SNM increases linearly until it reaches

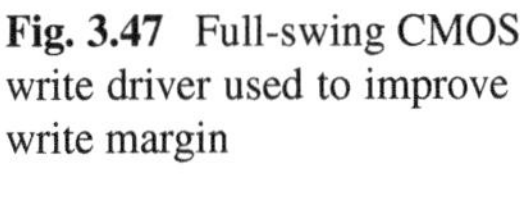

Fig. 3.47 Full-swing CMOS write driver used to improve write margin

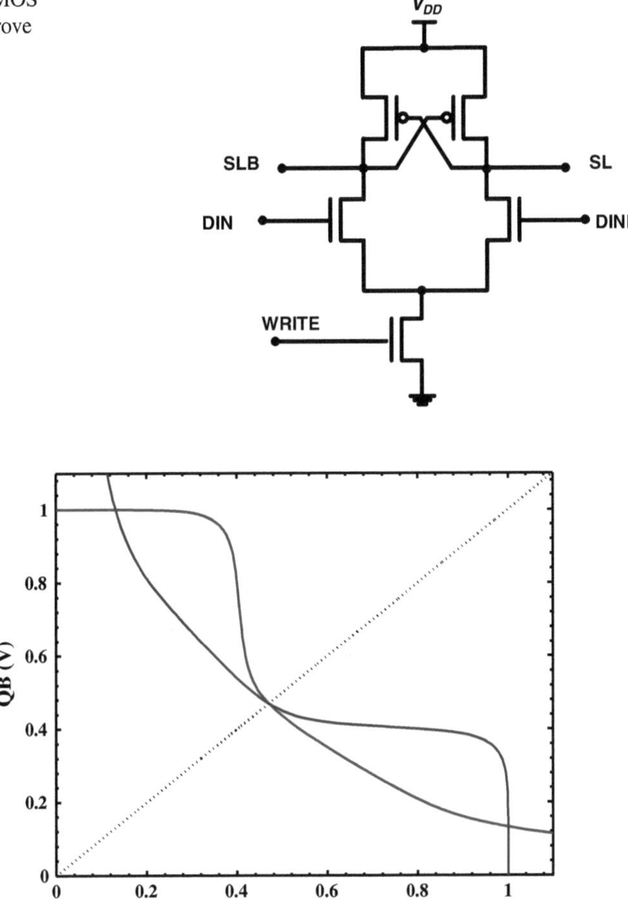

Fig. 3.48 SNM simulation results for nominal devices without WID variations

a maximum. Any further increase in ΔV_{BL} causes SNM to decrease significantly, which deteriorates cell read stability, agreeing with previous results [29, 39].

To evaluate the robustness of the proposed scheme in precisely controlling ΔV_{BL}, different process corners and post-layout RC extraction options were simulated, as shown in Table 3.3. ΔV_{BL} does not change significantly across different conditions (9–12 %), demonstrating the robustness of the proposed technique against PVT variations.

c

Figure 3.51 shows the process window curves (V_{th}), which are used to determine the operating limit of the memory accounting for 6σ of local variation coverage. In

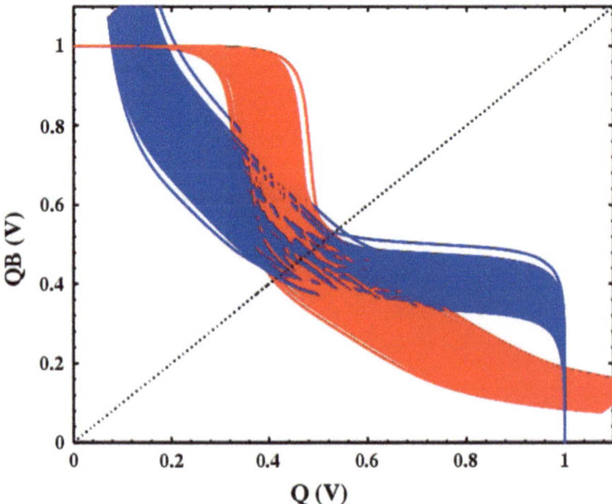

Fig. 3.49 SNM simulation results using Monte Carlo simulation for 1,000 MC runs

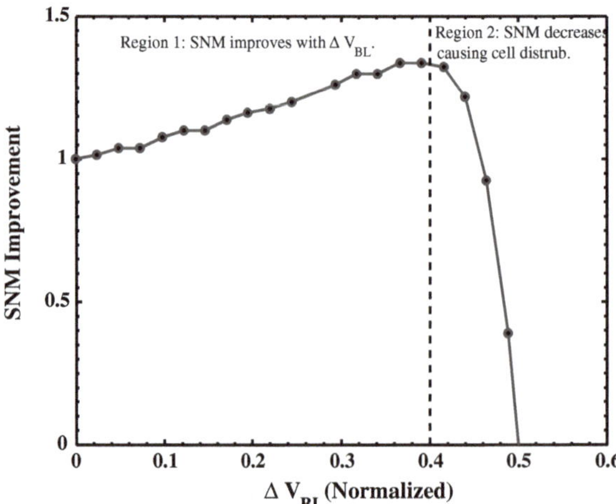

Fig. 3.50 SNM improvement versus ΔV_{BL} after charge sharing using the proposed technique

this simulation, the D2D variations are swept for NMOS and PMOS V_{th}. For each point of D2D variation, Monte Carlo simulation using WID variations is used to find the mean and sigma for SNM. We define the failure region as that where SNM reaches zero. Using the selective precharge technique (solid line), the operating window is expanded relative to the conventional approach. This increase in operation window reduces the failure probability by more than 100X.

To validate the improvements in cell stability using the proposed selective precharge technique, the designed 512 kb memory was fabricated in 45 nm tech-

Table 3.3 ΔV_{BL} for different conditions

Process	Slow	Slow	Nominal	Fast
Temp.	–40	125	25	–40
Parasitic C	Max	Max	Nominal	Min
ΔV_{BL}A	9.8 %	9.4 %	11.1 %	12 %

[A] Normalized to V_{DD}.

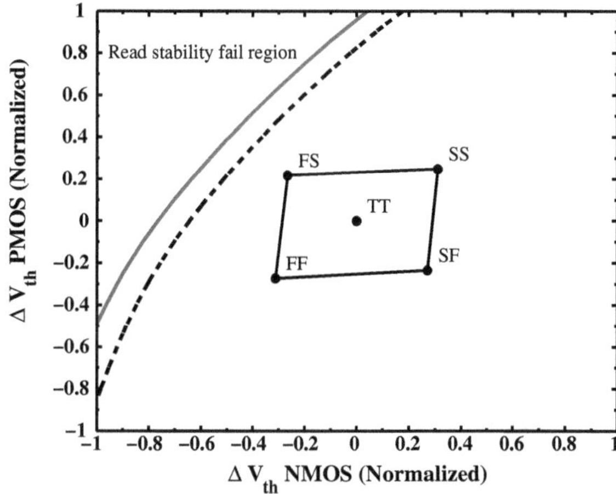

Fig. 3.51 Read stability operating window for selective precharge (*solid line*) compared to the conventional approach (*dotted line*). Simulation accounts for 6σ of local variations

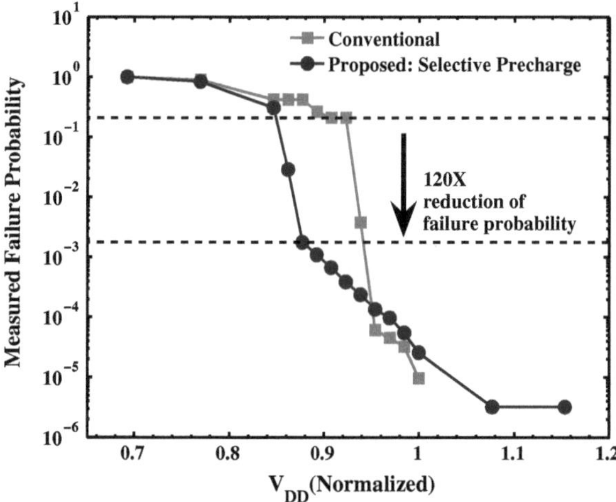

Fig. 3.52 Measured failure probability for the fabricated 512 kb memory for the proposed technique (selective precharge) and the conventional approach

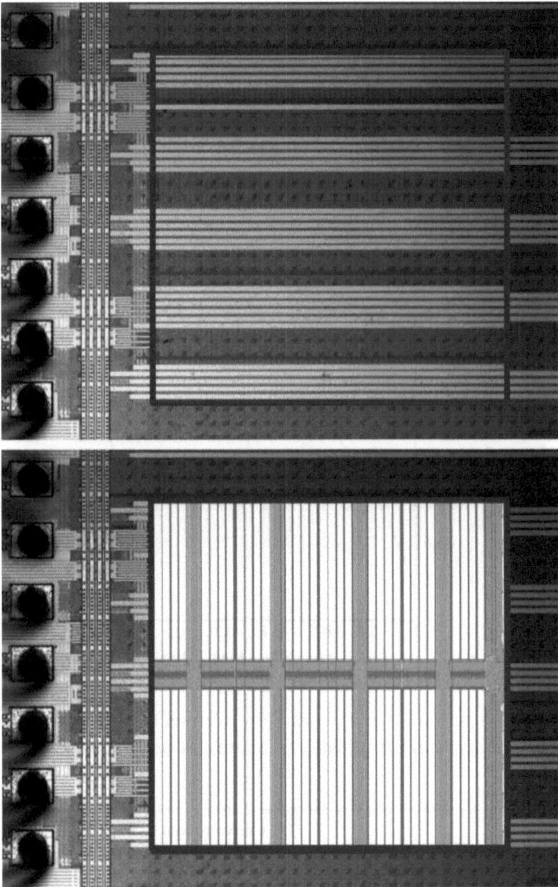

Fig. 3.53 Chip micrograph for the fabricated 512 kb memory in 45 nm technology. Upper figure shows the location of the memory and the lower overlays the memory layout

nology. Figure 3.53 shows the fabricated test chip micrograph. Measured results for cell failure probability using the conventional and the proposed technique are shown in Fig. 3.52, which shows that the proposed technique reduces the failure probability by more than 120X, validating the improvement in bitcell read stability.

The proposed technique has a small area overhead ($<2\,\%$) and shows strong robustness against process variations. In addition, it requires only one supply voltage, eliminating the need of level-shifters that cause large area and speed penalty. Moreover, the timing generation is simple since it re-uses timing signals available in SRAM design. Finally, the technique improves the memory speed, bitcell stability, and operating window, demonstrating its effectiveness.

3.9 Summary

The increase in local variations in nanometer technologies strongly affects SRAM cell stability. Various metrics used to analyze write and read stability were discussed. Detailed overview of the state-of-the art assist techniques and their impact on conventional SRAM design approaches was presented. As a case study, the implementation details of a single supply read assist technique were discussed. The proposed technique, selective precharge, allows precharging different parts of the bitlines to V_{DD} and GND and uses charge sharing to precisely control the bitline voltage which increases the bitcell stability. In addition to improving SNM, the proposed technique also improves memory access time. A 512 kb memory was designed to demonstrate the effectiveness of this technique in an industrial 45 nm technology. The technique significantly improves read stability, and provides high robustness against process variations.

References

1. K. Itoh, M. Horiguchi, M. Yamaoka, Low-voltage limitations of memory-rich nano-scale CMOS LSIs, in *33rd European Solid State Circuits Conference, 2007. ESSCIRC*, pp. 68–75, 11–13 Sept. 2007
2. A. Bhavnagarwala, S. Kosonocky, Y. Chan, K. Stawiasz, U. Srinivasan, S. Kowalczyk, M. Ziegler, A sub-600 mv, fluctuation tolerant 65 nm CMOS SRAM array with dynamic cell biasing, in *Proceedings of IEEE Symposium on VLSI Circuits*, pp. 78–79, 2007
3. S. Mukhopadhyay, H. Mahmoodi, K. Roy, Statistical design and optimization of SRAM cell for yield enhancement, in *Proceedings of International conference on Computer Aided Design*, pp. 10–13, 2004
4. E. Grossar, M. Stucchi, K. Maex, W. Dehaene, Read stability and write-ability analysis of SRAM cells for nanometer technologies. IEEE J. Solid-State Circ. **41**(11), 2577–2588 (2006)
5. M. Yamaoka, N. Maeda, Y. Shinozaki, Y. Shimazaki, K. Nii, S. Shimada, K. Yanagisawa, T. Kawahara, Low-power embedded SRAM modules with expanded margins for writing, in *Proceedings of the International Solid-State Circuits Conference ISSCC*, vol. 1, pp. 480–611, 2005
6. K. Takeda, H. Ikeda, Y. Hagihara, M. Nomura, H. Kobatake, Redefinition of write margin for next-generation SRAM and write-margin monitoring circuit, in *Proceedings of the International Solid-State Circuits Conference ISSCC*, pp. 2602–2611, 2006
7. S. Mukhopadhyay, H. Mahmoodi, K. Roy, Modeling of failure probability and statistical design of SRAM array for yield enhancement in nanoscaled CMOS. Comput. Aided Des. Integr. Circ. Syst. IEEE Trans. **24**(12) PP. 1859–1880, Dec. 2005
8. W. Dong, P. Li, G. Huang, SRAM dynamic stability: Theory, variability and analysis, in *IEEE/ACM International Conference on Computer-Aided Design, ICCAD 2008*, pp. 378–385, Nov. 2008
9. Y. Zhang, P. Li, G.M. Huang, Separatrices in high-dimensional state space: system-theoretical tangent computation and application to SRAM dynamic stability analysis, in *Proceedings of the 47th Design Automation Conference*, ser. DAC ’10. (ACM, USA, 2010), pp. 567–572
10. M. Khellah, D. Khalil, D. Somasekhar, Y. Ismail, T. Karnik, V. De, Effect of power supply noise on SRAM dynamic stability, in *Proceedings of IEEE Symposium on VLSI Circuits*, pp. 76–77, June 2007

11. A. Bhavnagarwala, S. Kosonocky, C. Radens, K. Stawiasz, R. Mann, Q. Ye, K. Chin, Fluctuation limits and scaling opportunities for CMOS SRAM cells, in *Proceedings of the International Electron Devices Meeting (IEDM)*, pp. 659–662, 2005
12. J. Wang, S. Nalam, B. Calhoun, Analyzing static and dynamic write margin for nanometer srams, in *ACM/IEEE International Symposium on Low Power Electronics and Design (ISLPED)*, pp. 129–134, Aug 2008
13. K. Agarwal, S. Nassif, Statistical analysis of SRAM cell stability, in *DAC '06: Proceedings of the 43rd Annual Conference on Design Automation*, pp. 57–62, 2006
14. G. Huang, W. Dong, Y. Ho, P. Li, Tracing SRAM separatrix for dynamic noise margin analysis under device mismatch, in *Behavioral Modeling and Simulation Workshop, BMAS 2007. IEEE International*, pp. 6–10, Sept. 2007
15. S. Nalam, V. Chandra, R. Aitken, B. Calhoun, Dynamic write limited minimum operating voltage for nanoscale SRAMs, in *Design, Automation Test in Europe Conference Exhibition (DATE)*, pp. 1–6, March 2011
16. R. Joshi, R. Kanj, S. Nassif, D. Plass, Y. Chan, C.-T. Chuang, Statistical exploration of the dual supply voltage space of a 65nm PD/SOI CMOS SRAM cell, in *Solid-State Device Research Conference, 2006. ESSDERC 2006. Proceeding of the 36th European*, pp. 315–318, Sept. 2006
17. M. Yamaoka, K. Osada, T. Kawahara, A cell-activation-time controlled SRAM for low-voltage operation in DVFS SoCs using dynamic stability analysis, *ESSCIRC: 34th European Solid State Circuits Conference*, pp. 286–289, Sept. 2008
18. S.O. Toh, Z. Guo, B. Nikolić, Dynamic SRAM stability characterization in 45nm CMOS, in *IEEE Symposium on VLSI Circuits (VLSIC)*, pp. 35–36, June 2010
19. S. Ikeda, Y. Yoshida, K. Ishibashi, Y. Mitsui, Failure analysis of 6T SRAM on low-voltage and high-frequency operation. IEEE Trans. Electron Devices **50**, 1270–1276 (2003)
20. R. Heald, P. Wang, Variability in sub-100 nm SRAM designs, in *Proceedings of International conference on Computer Aided Design*, pp. 347–352, 2004
21. E. Seevinck, F. List, J. Lohstroh, Static-noise margin analysis of MOS SRAM cells. IEEE J. Solid-State Circ. **22**(5), 748–754 (1987)
22. K. Nii, M. Yabuuchi, Y. Tsukamoto, S. Ohbayashi, S. Imaoka, H. Makino, Y. Yamagami, S. Ishikura, T. Terano, T. Oashi, K. Hashimoto, A. Sebe, S. Okazaki, K. Satomi, H. Akamatsu, H. Shinohara, A 45-nm bulk CMOS embedded SRAM with improved immunity against process and temperature variations. IEEE J. Solid-State Circ. **43**(1), 180–191 (2008)
23. C. Wann, R. Wong, D. Frank, R. Mann, S.-B. Ko, P. Croce, D. Lea, D. Hoyniak, Y.-M. Lee, J. Toomey, M. Weybright, J. Sudijono, SRAM cell design for stability methodology, in *IEEE VLSI-TSA International Symposium on VLSI Technology (VLSI-TSA-Tech)*, pp. 21–22, April 2005
24. A. Kawasumi, T. Yabe, Y. Takeyama, O. Hirabayashi, K. Kushida, A. Tohata, T. Sasaki, A. Katayama, G. Fukano, Y. Fujimura, N. Otsuka, A single-power-supply 0.7 V 1 GHz 45 nm SRAM with an asymmetrical unit-β-ratio memory cell, in *Solid-State Circuits Conference, ISSCC 2008. Digest of Technical Papers. IEEE. International*, pp. 382–622, Feb. 2008
25. M. Sharifkhani, M. Sachdev, SRAM cell stability: A dynamic perspective. IEEE J. Solid-State Circ. **44**(2), 609–619 (2009)
26. M. Sinangil, H. Mair, A. Chandrakasan, A 28 nm high-density 6T SRAM with optimized peripheral-assist circuits for operation down to 0.6 V, in *Solid-State Circuits Conference Digest of Technical Papers (ISSCC), IEEE. International*, pp. 260–262, Feb. 2011
27. B. Zhang, A. Arapostathis, S. Nassif, M. Orshansky, Analytical modeling of SRAM dynamic stability, in *IEEE/ACM International Conference on Computer-Aided Design, ICCAD '06*, pp. 315–322, Nov. 2006
28. M. Wieckowski, D. Sylvester, D. Blaauw, V. Chandra, S. Idgunji, C. Pietrzyk, R. Aitken, A black box method for stability analysis of arbitrary SRAM cell structures, in *Design, Automation Test in Europe Conference Exhibition (DATE)*, pp. 795–800, March 2010
29. M. Khellah, Y. Ye, N. Kim, D. Somasekhar, G. Pandya, A. Farhang, K. Zhang, C. Webb, V. De, Wordline and bitline pulsing schemes for improving SRAM cell stability in low Vcc 65 nm CMOS designs, in *Proceedings of IEEE Symposium on VLSI Circuits*, pp. 9–10, 2006

30. M. Khellah, D. Somasekhar, Y. Ye, N.S. Kim, J. Howard, G. Ruhl, M. Sunna, J. Tschanz, N. Borkar, F. Hamzaoglu, G. Pandya, A. Farhang, K. Zhang, V. De, A 256-kb dual-V_{CC} SRAM building block in 65-nm CMOS process with actively clamped sleep transistor. IEEE J. Solid-State Circ. **42**(1), 233–242 (2007)
31. L. Chang, D. Fried, J. Hergenrother, J. Sleight, R. Dennard, R. Montoye, L. Sekaric, S. McNab, A. Topol, C. Adams, K. Guarini, W. Haensch, Stable SRAM cell design for the 32 nm node and beyond, in *Symposium on VLSI Technology, 2005. Digest of Technical Papers*, pp. 128–129, June 2005
32. K. Takeda, Y. Hagihara, Y. Aimoto, M. Nomura, Y. Nakazawa, T. Ishii, H. Kobatake, A read-static-noise-margin-free SRAM cell for low-vdd and high-speed applications. IEEE J. Solid-State Circ. **41**(1), 113–121 (2006)
33. I.J. Chang, J.-J. Kim, S. Park, K. Roy, A 32 kb 10 T sub-threshold SRAM array with bit-interleaving and differential read scheme in 90 nm CMOS. IEEE J. Solid-State Circ. **44**(2), 650–658 (2009)
34. S. Jain, S. Khare, S. Yada, V. Ambili, P. Salihundam, S. Ramani, S. Muthukumar, M. Srinivasan, A. Kumar, S.K. Gb, R. Ramanarayanan, V. Erraguntla, J. Howard, S. Vangal, S. Dighe, G. Ruhl, P. Aseron, H. Wilson, N. Borkar, V. De, S. Borkar, A 280 mV-to-1.2 V wide-operating-range IA-32 processor in 32 nm CMOS, in *IEEE. International on Solid-State Circuits Conference Digest of Technical Papers (ISSCC)*, pp. 66–68, Feb. 2012
35. K. Zhang, U. Bhattacharya, Z. Chen, F. Hamzaoglu, D. Murray, N. Vallepalli, Y. Wang, B. Zheng, M. Bohr, A 3-GHz 70-Mb SRAM in 65-nm cmos technology with integrated column-based dynamic power supply. IEEE J. Solid-State Circ. **41**(1), 146–151 (2006)
36. R.W. Mann, J. Wang, S. Nalam, S. Khanna, G. Braceras, H. Pilo, B.H. Calhoun, Impact of circuit assist methods on margin and performance in 6T SRAM. Solid-State Electron. **54**(11), 1398–1407 (2010)
37. H. Yamauchi, A discussion on SRAM circuit design trend in deeper nanometer-scale technologies. IEEE Trans. Very Large Scale Integr. (VLSI) Syst. **18**(5), 763–774 (2010)
38. M. Sinangil, N. Verma, A. Chandrakasan, A reconfigurable 65 nm SRAM achieving voltage scalability from 0.25 to 1.2 V and performance scalability from 20kHz to 200 MHz, in *Solid-State Circuits Conference, 2008. ESSCIRC 2008. 34th European*, pp. 282–285, Sept. 2008
39. B. Campbell, J. Burnette, N. Javarappa, V. von Kaenel, Power-efficient dual-supply 64 kB L1 caches in a 65 nm CMOS technology, in *Proceedings of IEEE Custom Integrated Circuits Conference*, pp. 729–732, 2007
40. T. Suzuki, H. Yamauchi, K. Satomi, H. Akamatsu, A stable SRAM mitigating cell-margin asymmetricity with a disturb-free biasing scheme, in *Proceedings of IEEE Custom Integrated Circuits conference*, pp. 233–236, 2007
41. R. Joshi, R. Houle, D. Rodko, P. Patel, W. Huott, R. Franch, Y. Chan, D. Plass, S. Wilson, S. Wu, and R. Kanj, A high performance 2.4 Mb L1 and L2 cache compatible 45 nm SRAM with yield improvement capabilities, in *Proceedings of IEEE Symposium on VLSI Circuits*, pp. 208–209, June 2008
42. J. Pille, C. Adams, T. Christensen, S. Cottier, S. Ehrenreich, T. Kono, D. Nelson, O. Takahashi, S. Tokito, O. Torreiter, O. Wagner, D. Wendel, Implementation of the cell broadband engine in a 65 nm SOI technology featuring dual-supply SRAM arrays supporting 6 GHz at 1.3 V, in *Proceedings of the International Solid-State Circuits Conference ISSCC*, pp. 322–606, 11–15 Feb. 2007
43. K. Zhang, U. Bhattacharya, Z. Chen, F. Hamzaoglu, D. Murray, N. Vallepalli, Y. Wang, B. Zheng, M. Bohr, A 3-GHz 70 MB SRAM in 65 nm CMOS technology with integrated column-based dynamic power supply, in *Proceedings of the International Solid-State Circuits Conference ISSCC*, vol. 1, pp. 474–611, 10–10 Feb. 2005
44. M. Yamaoka, T. Kawahara, Operating-margin-improved SRAM with column-at-a-time body-bias control technique, in *33rd European Solid State Circuits Conference, 2007. ESSCIRC*, pp. 396–399, 11–13 Sept. 2007
45. O. Hirabayashi, A. Kawasumi, A. Suzuki, Y. Takeyama, K. Kushida, T. Sasaki, A. Katayama, G. Fukano, Y. Fujimura, T. Nakazato, Y. Shizuki, N. Kushiyama, T. Yabe, A process-

variation-tolerant dual-power-supply SRAM with 0.179 um2 cell in 40 nm CMOS using level-programmable wordline driver, in *IEEE International on Solid-State Circuits Conference - Digest of Technical Papers, 2009. ISSCC 2009*, pp. 458–459,459a, Feb. 2009

46. F. shi Lai, C.-F. Lee, On-chip voltage down converter to improve SRAM read/write margin and static power for sub-nano CMOS technology. IEEE J. Solid-State Circ. **42**(9), 2061–2070 (2007)
47. Y. Hirano, M. Tsujiuchi, K. Ishikawa, H. Shinohara, T. Terada, Y. Maki, T. Iwamatsu, K. Eikyu, T. Uchida, S. Obayashi, K. Nii, Y. Tsukamoto, M. Yabuuchi, T. Ipposhi, H. Oda, Y. Inoue, A robust SOI SRAM architecture by using advanced ABC technology for 32 nm node and beyond LSTP devices, in *Proceedings of IEEE Symposium on VLSI Technology*, pp. 78–79, June 2007
48. Y. Morita, H. Fujiwara, H. Noguchi, K. Kawakami, J. Miyakoshi, S. Mikami, K. Nii, H. Kawaguchi, and M. Yoshimoto, A Vth-Variation-Tolerant SRAM with 0.3-V minimum operation voltage for memory-rich SoC under DVS environment, in *Proceedings of IEEE Symposium on VLSI Circuits*, pp. 13–14, 2006
49. H. Pilo, C. Barwin, G. Braceras, C. Browning, S. Lamphier, F. Towler, An SRAM design in 65-nm technology node featuring read and write-assist circuits to expand operating voltage. IEEE J. Solid-State Circ. **42**(4), 813–819 (2007)
50. Y.H. Chen, G. Chan, S.Y. Chou, H.-Y. Pan, J.-J. Wu, R. Lee, H. Liao, H. Yamauchi, A 0.6 V dual-rail compiler SRAM design on 45 nm CMOS technology with adaptive SRAM power for lower VDD_{min} VLSIs. IEEE J. Solid-State Circ. **44**(4), 1209–1215 (2009)
51. S. Ohbayashi, M. Yabuuchi, K. Nii, Y. Tsukamoto, S. Imaoka, Y. Oda, T. Yoshihara, M. Igarashi, M. Takeuchi, H. Kawashima, Y. Yamaguchi, K. Tsukamoto, M. Inuishi, H. Makino, K. Ishibashi, H. Shinohara, A 65-nm SoC embedded 6T-SRAM designed for manufacturability with read and write operation stabilizing circuits. IEEE J. Solid-State Circ. **42**(4), 820–829 (2007)
52. A. Raychowdhury, B. Geuskens, J. Kulkarni, J. Tschanz, K. Bowman, T. Karnik, S.-L. Lu, V. De, M. Khellah, PVT-and-aging adaptive wordline boosting for 8T SRAM power reduction, in *IEEE International on Solid-State Circuits Conference Digest of Technical Papers (ISSCC)*, pp. 352–353, Feb. 2010
53. M. Yamaoka, N. Maeda, Y. Shinozaki, Y. Shimazaki, K. Nii, S. Shimada, K. Yanagisawa, T. Kawahara, 90-nm process-variation adaptive embedded SRAM modules with power-line-floating write technique. IEEE J. Solid-State Circ. **41**(3), 705–711 (2006)
54. J. Kulkarni, B. Geuskens, T. Karnik, M. Khellah, J. Tschanz, V. De, Capacitive-coupling wordline boosting with self-induced VCC collapse for write VMIN reduction in 22-nm 8T SRAM, in *IEEE International on Solid-State Circuits Conference Digest of Technical Papers (ISSCC)*, pp. 234–236, Feb. 2012
55. E. Karl, Y. Wang, Y.-G. Ng, Z. Guo, F. Hamzaoglu, U. Bhattacharya, K. Zhang, K. Mistry, M. Bohr, A 4.6 GHz 162 Mb SRAM design in 22 nm tri-gate CMOS technology with integrated active VMIN-enhancing assist circuitry, in *IEEE International on Solid-State Circuits Conference Digest of Technical Papers (ISSCC)*, pp. 230–232, Feb. 2012
56. S. Damaraju, V. George, S. Jahagirdar, T. Khondker, R. Milstrey, S. Sarkar, S. Siers, I. Stolero, A. Subbiah, A 22 nm IA multi-CPU and GPU system-on-chip, in *IEEE International on Solid-State Circuits Conference Digest of Technical Papers (ISSCC)*, pp. 56–57, Feb. 2012
57. M. Khellah, N.S. Kim, Y. Ye, D. Somasekhar, T. Karnik, N. Borkar, F. Hamzaoglu, T. Coan, Y. Wang, K. Zhang, C. Webb, V. De, PVT-variations and supply-noise tolerant 45 nm dense cache arrays with diffusion-notch-free (DNF) 6T SRAM cells and dynamic multi-vcc circuits, in *2008 IEEE Symposium on VLSI Circuits*, pp. 48–49, June 2008
58. B. Mohammad, M. Saint-Laurent, P. Bassett, J. Abraham, Cache design for low power and high yield, in *9th International Symposium on Quality Electronic Design, 2008. ISQED 2008*, pp. 103–107, March 2008
59. Y. Wang, E. Karl, M. Meterelliyoz, F. Hamzaoglu, Y.-G. Ng, S. Ghosh, L. Wei, U. Bhattacharya, K. Zhang, Dynamic behavior of SRAM data retention and a novel transient voltage collapse technique for 0.6 V 32 nm LP SRAM, in *IEEE International on Electron Devices Meeting (IEDM)*, pp. 32.1.1–32.1.4, Dec. 2011

60. K. Nii, M. Yabuuchi, Y. Tsukamoto, S. Ohbayashi, Y. Oda, K. Usui, T. Kawamura, N. Tsuboi, T. Iwasaki, K. Hashimoto, H. Makino, H. Shinohara, A 45-nm single-port and dual-port SRAM family with robust read/write stabilizing circuitry under DVFS environment, in *IEEE Symposium on VLSI Circuits*, pp. 212–213, June 2008
61. Y. Fujimura, O. Hirabayashi, T. Sasaki, A. Suzuki, A. Kawasumi, Y. Takeyama, K. Kushida, G. Fukano, A. Katayama, Y. Niki, T. Yabe, A configurable SRAM with constant-negative-level write buffer for low-voltage operation with 0.149um2 cell in 32 nm high-k metal-gate CMOS, in *IEEE International on Solid-State Circuits Conference Digest of Technical Papers (ISSCC)*, pp. 348–349, Feb. 2010
62. N. Shibata, H. Kiya, S. Kurita, H. Okamoto, M. Tan'no, T. Douseki, A 0.5-V 25-MHz 1-mW 256-kb MTCMOS/SOI SRAM for solar-power-operated portable personal digital equipment - sure write operation by using step-down negatively overdriven bitline scheme. IEEE J. Solid-State Circ. **41**(3), pp. 728–742, March 2006
63. S. Mukhopadhyay, R. Rao, J.-J. Kim, C.-T. Chuang, SRAM write-ability improvement with transient negative bit-line voltage. IEEE Trans. Very Large Scale Integr. (VLSI) Syst. **19**(1), 24–32 (2011)
64. H. Pilo, I. Arsovski, K. Batson, G. Braceras, J. Gabric, R. Houle, S. Lamphier, F. Pavlik, A. Seferagic, L.-Y. Chen, S.-B. Ko, C. Radens, A 64 Mb SRAM in 32 nm High-k metal-gate SOI technology with 0.7 V operation enabled by stability, write-ability and read-ability enhancements, in *IEEE International on Solid-State Circuits Conference Digest of Technical Papers (ISSCC)*, pp. 254–256, Feb. 2011
65. H. Nho, P. Kolar, F. Hamzaoglu, Y. Wang, E. Karl, Y.-G. Ng, U. Bhattacharya, K. Zhang, A 32 nm High-k metal gate SRAM with adaptive dynamic stability enhancement for low-voltage operation, in *IEEE International on Solid-State Circuits Conference Digest of Technical Papers (ISSCC)*, pp. 346–347, Feb. 2010
66. M. Yabuuchi, K. Nii, Y. Tsukamoto, S. Ohbayashi, S. Imaoka, H. Makino, Y. Yamagami, S. lshikura, T. Terano, T. Oashi, K. Hashimoto, A. Sebe, G. Okazaki, K. Satomi, H. Akamatsu, H. Shinohara, A 45 nm low-standby-power embedded SRAM with improved immunity against process and temperature variations, in *Proceedings of the International Solid-State Circuits Conference ISSCC*, pp. 326–606, 11–15 Feb. 2007
67. M. H. Abu-Rahma, M. Anis, S.S. Yoon, A robust single supply voltage SRAM read assist technique using selective precharge, in *Proceedings of the 34th European Solid State Circuits Conference ESSCIRC*, pp. 234–237, 2008
68. B. Wicht, T. Nirschl, D. Schmitt-Landsiedel, Yield and speed optimization of a latch-type voltage sense amplifier. IEEE J. Solid-State Circ **39**(7), 1148–1158 (2004)

Chapter 4
Reducing SRAM Power Using Fine-Grained Wordline Pulse Width Control

4.1 Introduction

Modeling and optimizing memory yield at higher levels of design abstraction (i.e., system and architecture) provides enhanced variation tolerance capabilities compared to improvements due to circuit changes alone. Conventional architecture approaches such as memory redundancy and ECC which were previously used to deal with hard defects are now used to reduce SRAM $V_{\min}$ [1, 2]. Recently, adaptive architectures that incorporate on-chip sensors to automatically adjust the supply voltage [3], body bias [4, 5], WL voltage level [6, 7], or WL internal timing have been proposed [8].

A self-V_{DD} tuning scheme that automatically adjusts memory supply voltage to near its functional $V_{\min}$ has been used to reduce power consumption while meeting the performance requirement [3]. Figure 4.1 shows an overview of the system, which uses an on-chip voltage regulator to control the memory voltage, an all-digital PLL to generate the required clock frequency, and a BIST to test the memory at a given supply voltage and operating frequency. Similar self-adjusting architectures use adaptive body bias: an on-chip monitor determines the amount of global variation, and the system selects the optimal body bias voltage to reduce the memory failure probability ABB [4, 5].

System-specific memory optimization has also been proposed to improve memory yield at low voltage. For many applications, the system can accept and possibly correct memory errors. Therefore, co-designing the system and the memory simultaneously achieves higher tolerance against process variations without significant impact on the overall system performance [9]. Examples of these systems include video decoders [9, 10], wireless systems [9], and processor caches [2, 11]. Specifically, variation-tolerant cache architectures have been proposed to improve memory yield with little impact on performance [2, 11]. The architecture dynamically detects and replaces faulty cells using dynamic cache resizing. The technique enhances yield more than static repair techniques such as redundancy and ECC, which makes the architecture attractive for low voltage operation [2, 11].

M. H. Abu-Rahma and M. Anis, *Nanometer Variation-Tolerant SRAM*,
DOI: 10.1007/978-1-4614-1749-1_4,

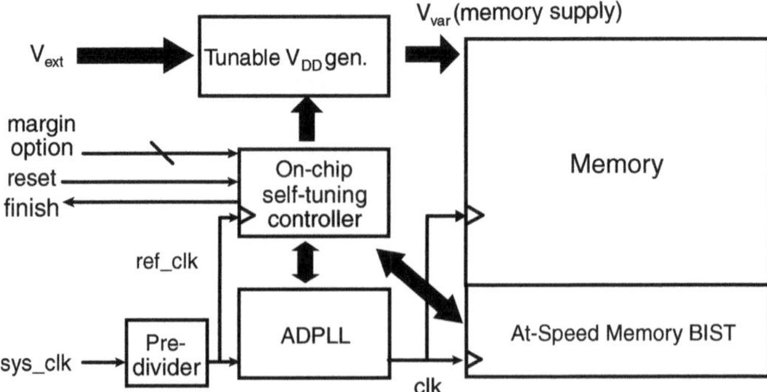

Fig. 4.1 A system that automatically tunes the memory supply voltage to reduce the power consumption [3]

Another example of system-specific memory design has been proposed in the design of a variation-tolerant architecture for MPEG-4 video processors [10]. The system reduces power consumption with only slight degradation in quality. The technique relies on the fact that human visual system is sensitive mainly to higher order bits of luminance pixels in the video data. This property of video data is exploited to implement a storage policy in which higher order bits are stored in 8T bitcells (which have lower $V_{\min}$ than 6T bitcells, but with much larger area overhead), while lower order bits are stored in conventional 6T bitcells (small area). Because of this setup, as supply voltage is reduced, the information of the critical higher order bits stay stored in the 8T bitcells. This technique facilitates aggressive voltage scaling and power reduction with negligible video quality degradation [10].

In the next sections, we look at a new architectural technique to reduce the memory power consumption in the presence of large process variations [8].

4.2 Motivation

With technology scaling, the requirements of higher density and lower power embedded SRAM are increasing exponentially. More than 90 % of die area in future System-on-Chip (SoC) will be occupied by SRAM [12]. This is driven by the high demand for low-power mobile systems, which integrate a wide range of functionality such as digital cameras, 3D graphics, MP3 players, and other applications. In the mean time, random variations are increasing significantly with technology scaling as discussed in Chap. 2. The variations in V_{th} are inversely proportional to the square root of the device area [13]. Therefore, SRAM bitcells experience the largest random variations on a chip, as bitcell transistors are typically the smallest devices in a technology [8, 14–17].

Embedded SRAMs usually dominate the SoC silicon area, and their power consumption (both dynamic and static) is a considerable portion of the total power consumption of an SoC. Moreover, SRAM yield can dominate the overall chip yield. However, to achieve high yield, memory power consumption and speed are negatively impacted. The stringent requirements of high yield and low power consumption requires combining circuit and architectural techniques in future SRAM designs.

SRAM array switching power consumption is considered one of the largest components of power in high density memories [18, 19]. This results from the fact that requirements for high area efficiency forces SRAM designers to maximize the array sizes. Figure 4.2 shows dynamic power consumption for read operation versus wordline pulse width T_{wl} for a 512 kb memory macro in an industrial 65 nm technology. Power consumption results are extrapolated to $T_{\mathrm{wl}} = 0$ to estimate the component of switching power due to the peripheral circuits. For normal operating conditions, array power consumption is more than 60 % of read power. Therefore, it is important to reduce the array switching power due to its strong impact on the memory's total power as well as the SoC's power.

Several circuit techniques have been proposed to reduce SRAM array switching power consumption by reducing wordline pulse width. One of the most common techniques to control T_{wl} is using a bitcell replica path, which reduces the bitline differential and, therefore, lowers power consumption [15, 17, 20–22]. Replica path (e.g., self-timed) techniques provide a simple approach of process tracking for global variations (interdie or systematic within-die) as well as environmental variations (voltage and temperature). However, these circuit techniques are not efficient when memory bitcells experience large random variation, since circuit techniques cannot adapt to random variations. Therefore, their effectiveness decreases with process scaling, so larger design margins are used, which increases power consumption due to larger T_{wl}. To reduce the loss due to excessive margining, circuits, and architecture must be designed together to reduce power and manage variability. Higher levels of design abstraction can have better variation-tolerance capabilities because the impact of random variation can be easily measured more accurately at that level. Therefore, combining architecture techniques with circuit level can help adapt the memory to random variation and reduce power consumption [14, 23, 24].

4.3 Yield and Power Tradeoff

Due to random variation affecting SRAM bitcells, memory yield, and power consumption are tightly. To achieve high yield, read access failures should be minimized. Read access yield is defined as the probability of correct read operation. In this chapter, we use the term yield to refer to read access yield, which does not include read stability, writeability or retention failures. In read operation, the selected wordline is activated for a period of time to allow the bitlines to discharge. The wordline activation time (T_{wl}) is a critical parameter for memory design since it affects the

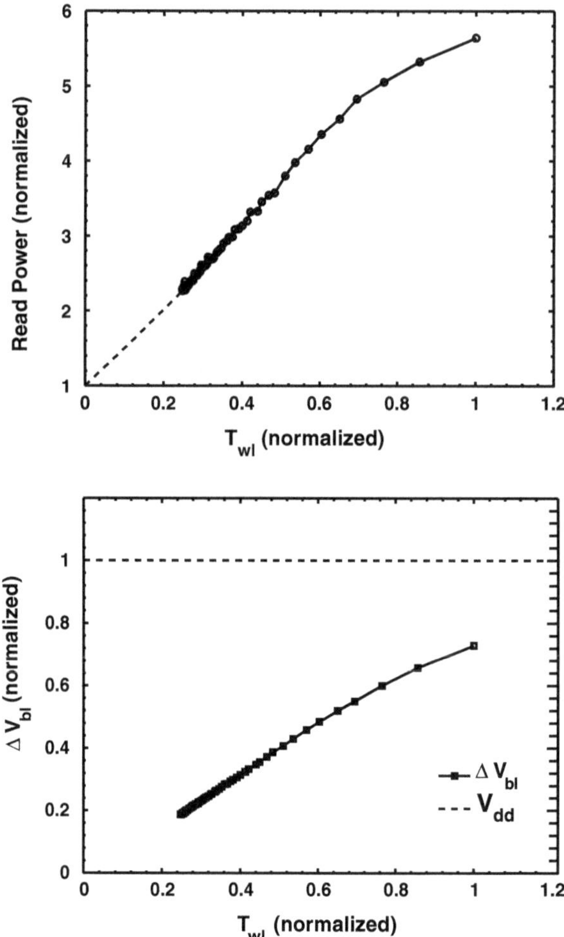

Fig. 4.2 Memory read power and bitline differential versus T_{wl} for a 512 kb memory in 65 nm technology

memory speed (access time) as well as memory power. To reduce read access failures, the wordline pulse T_{wl} should be large enough to guarantee adequate bitline voltage, which can be sensed via a sense amplifier.

The total power consumption for a memory in a read or write cycle can be expressed as:

$$P_{\mathrm{mem}} = P_{\mathrm{leak}} + P_{s_{\mathrm{array}}} + P_{s_{\mathrm{peri}}} \tag{4.1}$$

where P_{leak} is the total leakage power from the array and the peripheral circuitry, $P_{s_{\mathrm{array}}}$ and $P_{s_{\mathrm{peri}}}$ are the switching power from the array and the peripheral circuitry, respectively (as shown in Fig. 4.3).

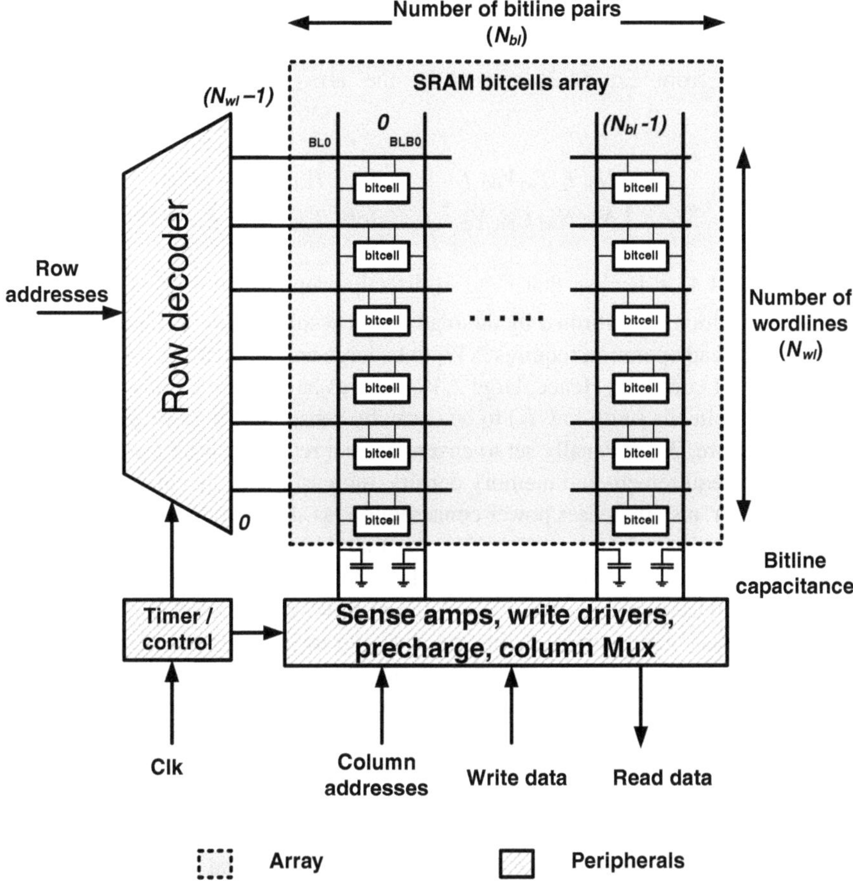

Fig. 4.3 Typical SRAM architecture

In a read access, the array switching power can be calculated as:

$$P_{s_{\text{array}}} = N_{\text{bl}}\, N_{\text{wl}}\, C_{\text{bit}}\, \Delta V_{\text{bl}}\, V_{\text{dd}}\, f \tag{4.2}$$

where N_{bl} are N_{wl} are the number of bitlines and wordlines in a memory bank, respectively. C_{bit} is the bitline capacitance per bitcell, ΔV_{bl} is the bitline differential voltage in read access (used to sense the bitcell's stored value), V_{dd} is the supply voltage, and f is the operating frequency.

ΔV_{bl} can be calculated as:

$$\Delta V_{\text{bl}} \simeq \begin{cases} \frac{I_c\, T_{\text{wl}}}{N_{\text{wl}}\, C_{\text{bit}}} & \text{for } \; T_{\text{wl}} \leq \frac{V_{\text{dd}}\, N_{\text{wl}}\, C_{\text{bit}}}{I_c} \\ V_{\text{dd}} & \text{for } \; T_{\text{wl}} > \frac{V_{\text{dd}}\, N_{\text{wl}}\, C_{\text{bit}}}{I_c} \end{cases} \tag{4.3}$$

where I_c is the bitcell read current. ΔV_{bl} can be approximated assuming linear dependence on T_{wl}, for the range of T_{wl} where $\Delta V_{\text{bl}} < V_{\text{dd}}$, as shown in Fig. 4.2.

Therefore, from Eqs. (4.2) and (4.3), the array switching power can be computed as:

$$P_{s_{\text{array}}} \simeq \begin{cases} N_{\text{bl}}\, I_c\, T_{\text{wl}} V_{\text{dd}}\, f & \text{for } T_{\text{wl}} \leq \frac{V_{\text{dd}} N_{\text{wl}} C_{\text{bit}}}{I_c} \\ N_{\text{bl}}\, N_{\text{wl}}\, C_{\text{bit}}\, V_{\text{dd}}{}^2\, f & \text{for } T_{\text{wl}} > \frac{V_{\text{dd}} N_{\text{wl}} C_{\text{bit}}}{I_c} \end{cases} \tag{4.4}$$

From Eq. (4.4), it is clear that $P_{s_{\text{array}}}$ is directly proportional to T_{wl} (when $T_{\text{wl}} \leq \frac{V_{\text{dd}} N_{\text{wl}} C_{\text{bit}}}{I_c}$), which is confirmed by the read power results shown in Fig. 4.2.

A correct read operation requires ΔV_{bl} to be large enough that the sense amplifier can measure it correctly. Hence, large ΔV_{bl} implies having sufficiently large T_{wl} to enable weak bitcells (with low I_c) to be correctly sensed, for a given yield requirement. Therefore, T_{wl} is usually set to ensure correct read operation for a given read access yield requirement and memory density. Increasing T_{wl} increases read access yield; however, also increases power consumption, as shown in Eq. (4.4). Moreover, T_{wl} has a direct impact on a memory's access time [15, 25].

Bitcell read current is strongly affected by the random Vth variations in the bitcell access device (pass gate) as well as the pull down device. Due to these variations, the bitcell read current I_c has been shown to follow a normal distribution $\mathcal{N} \sim (\mu_{I_c}, \sigma^2{}_{I_c})$ with a mean of μ_{I_c} and standard deviation of σ_{I_c} [15–17, 25, 26]. Measurements confirm that I_c follows a normal distribution up to 5σ value [27]. However, because of technology and supply scaling, and the increase of V_{th} variation, there are exceptions to this case. The large variations may cause the device operation region to change from strong-inversion to moderate-inversion. In the extreme case of variations, the device may operate in subthreshold region, hence I_c distribution will become log-normal due to the exponential dependence on V_{th} [28, 29]

Random WID variations also have strong impact on the sense amplifier (SA) offset voltage [15, 25, 30–32], which can affect the accuracy of read operation. In addition, systematic variations due to asymmetric layout can increase the SA input offset, which is why highly symmetric layouts are typically used for SA [15, 25]. Typically, the SA input offset due to random variations can be modeled using a normal distribution [31]. However, due to the complexity of statistically deriving a closed form yield relation that accounts for both I_c and SA offset, the SA input offset may be approximated according to worst-case approach [15, 16, 26] instead of the statistical approach [25].

To guarantee correct read operation, the following condition should be satisfied [15, 16, 26]:

$$T_{\text{wl,wc}} = \frac{\Delta V_{\text{bl,min}} N_{\text{wl}}\, C_{\text{bit}}}{\mu_{I_c}(1 - \frac{\sigma}{\mu}|_{I_c} N_\sigma)} \tag{4.5}$$

where $\Delta V_{\text{bl,min}}$ is the minimum required bitline differential voltage, which is a function of the sense amplifier input offset (typically $\Delta V_{\text{bl,min}} = 0.1\, V_{\text{dd}}$). μ_{I_c} is the mean bitcell read current, and $\frac{\sigma}{\mu}|_{I_c}$ is the relative variation in I_c. N_σ is the required design yield in terms of σ, which is related to the target yield and the memory density [15, 26], and can be computed as:

$$N_\sigma = \Phi^{-1}(Y_{\text{mem}}{}^{\frac{1}{N_{\text{bits}}}}) \tag{4.6}$$

where Φ^{-1} is the inverse standard normal cumulative distribution function, Y_{mem} is the memory target yield, and N_{bits} is the total number of bitcells in the memory. For example, for a 1 Mb memory, if the target read access yield is 95 %, then the required design coverage is $N_\sigma = 5.33$. Therefore, to achieve the same yield for larger memory densities, N_σ should be increased. From Eq. (4.5), this means that larger $T_{\text{wl,wc}}$ is required.

4.4 Derivation of the Probability Density Function of T_{wl} ($P_{T_{\text{wl}}}$)

The relation between $T_{\text{wl,wc}}$ and I_c is nonlinear as shown in Eqs. (4.3) and (4.5), therefore, the distribution of $T_{\text{wl,wc}}$ will not be normal distribution. To find the probability density function (PDF) of $T_{\text{wl,wc}}$, we use a one-to-one mapping from Eq. (4.5) as described below. Given a function $y = g(x)$, where the PDF of x is known, the PDF of y can be expressed as [33]:

$$f_Y(y) = \frac{f_X(x_1)}{|g'(x_1)|} + \cdots + \frac{f_X(x_n)}{|g'(x_n)|} \tag{4.7}$$

where $f_X(x)$ and $f_Y(y)$ are the PDFs of x and y, respectively. $g'(x)$ is the derivative of $g(x)$, and $x_1 \cdots x_n$ are the real roots of the equation $y = g(x)$. By using this equation to determine the PDF for T_{wl}, we begin with the relationship between T_{wl} and I_c from Eq. (4.5).

$$T_{\text{wl}}(I_c) = \frac{\Delta V_{\text{bl,min}} N_{\text{wl}}\, C_{\text{bit}}}{I_c} \tag{4.8}$$

and the derivative with respect to I_c becomes:

$$T'_{\text{wl}}(I_c) = -\frac{\Delta V_{\text{bl,min}} N_{\text{wl}}\, C_{\text{bit}}}{I_c{}^2} \tag{4.9}$$

Equation (4.8) has a single solution with $I_c = \frac{\Delta V_{\text{bl,min}} N_{\text{wl}}\, C_{\text{bit}}}{T_{\text{wl}}}$. Using Eqs. (4.7), (4.8), and (4.9), the probability density function of T_{wl} ($P_{T_{\text{wl}}}$) can be calculated as:

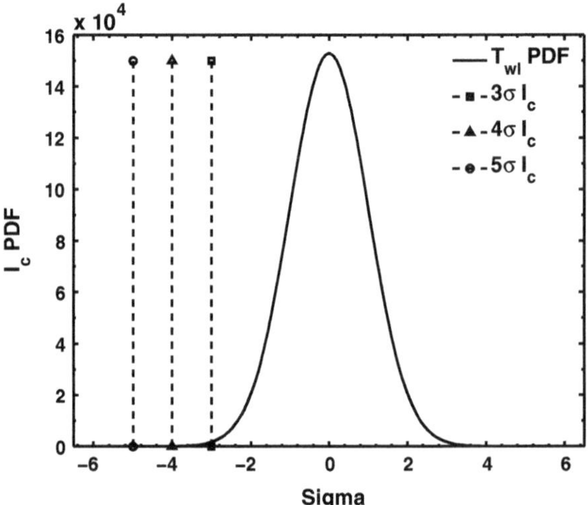

Fig. 4.4 I_c probability density function (PDF) showing 3, 4, and $5\sigma_{I_c}$: which correspond to different memory yield targets (assuming $\frac{\sigma}{\mu}|_{I_c} = 15\,\%$)

$$P_{T_{\rm wl}} = \frac{\varphi_{I_c}(I_c)}{|T'_{\rm wl}(I_c)|} \tag{4.10}$$

$$= \frac{\varphi_{I_c}\left(\frac{\Delta V_{\rm bl,min} N_{\rm wl}\, C_{\rm bit}}{T_{\rm wl}}\right)}{\left|-\frac{\Delta V_{\rm bl,min} N_{\rm wl}\, C_{\rm bit}}{\left(\frac{\Delta V_{\rm bl,min} N_{\rm wl}\, C_{\rm bit}}{T_{\rm wl}}\right)^2}\right|} \tag{4.11}$$

Therefore,

$$P_{T_{\rm wl}} = \frac{\Delta V_{\rm bl,min} N_{\rm wl}\, C_{\rm bit}}{{T_{\rm wl}}^2}\ \varphi_{I_c}\left(\frac{\Delta V_{\rm bl,min} N_{\rm wl}\, C_{\rm bit}}{T_{\rm wl}}\right) \quad \text{for} \quad T_{\rm wl} > 0 \tag{4.12}$$

where $\varphi_{I_c}()$ is the PDF for I_c, which follows a normal distribution $\mathcal{N} \sim (\mu_{I_c}, {\sigma^2}_{I_c})$.

Figures 4.4 and 4.5 show the distributions of I_c and $T_{\rm wl}$. Note that, the PDF of $T_{\rm wl}$ is not symmetric, but instead, is skewed toward larger $T_{\rm wl}$ values. Also, the 3, 4, and $5\sigma_{I_c}$ values are shown for I_c, and the corresponding $T_{\rm wl}$ values for these σ_{I_c} which show that $T_{\rm wl}$ is very sensitive to I_c variations. For $5\sigma_{I_c}$, $T_{\rm wl}$ increases four times compared to its nominal value (calculated using μ_{I_c}).

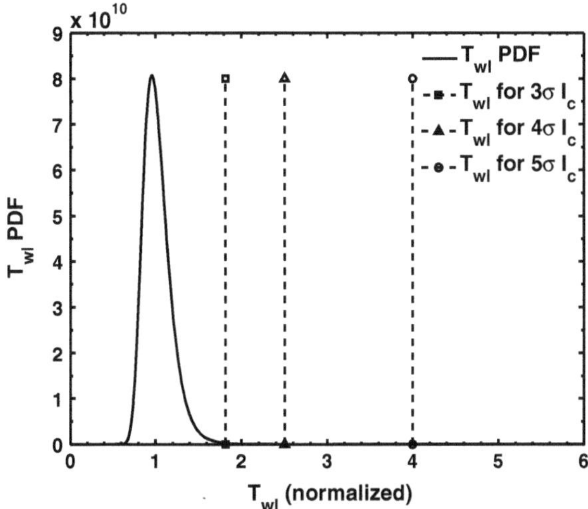

Fig. 4.5 T_{wl} Probability density function (PDF). Also shown are the points corresponding to 3, 4, $5\sigma_{I_c}$ (assuming $\frac{\sigma}{\mu}|_{I_c} = 15\,\%$) using Eq. (4.12)

Because of the skewed distribution of T_{wl}, large values of T_{wl} are required to ensure acceptable read access yield. Moreover, the higher N_σ required to achieve the same yield for larger densities significantly increases T_{wl} due to the nonlinear relation between I_c and T_{wl}, as in Eq. (4.6). Therefore, due to statistical variations in the bitcell, T_{wl} should be pessimistically large to achieve the target yield. This will increase the dynamic power for memories that do not have weak bitcells (have small variation of I_c). To reduce the power consumption, a post-silicon approach must be employed to compensate for the pessimism in determining T_{wl}.

4.5 Fine-Grained Wordline Pulse Width Control

As discussed in the previous section, read switching power increases significantly due to process variations, because at the design time we cannot determine which memories will have weak bitcells with low I_c. Therefore, a worst case approach is applied to determine $T_{\mathrm{wl,wc}}$ for a given memory density and yield requirement, as shown in Eq. (4.5).

Memory BIST is now an integral part of an SoC [12, 34]. Typically, a memory BIST engine is used to generate patterns to test a memory and locate memory failures. Based on the memory failure signature, the unit under test is either discarded or repaired using memory redundancy. To reduce SoC repair cost, an alternative repair approach is built-in self-repair (BISR) used to enable soft repair and reduce test and repair time [12, 34].

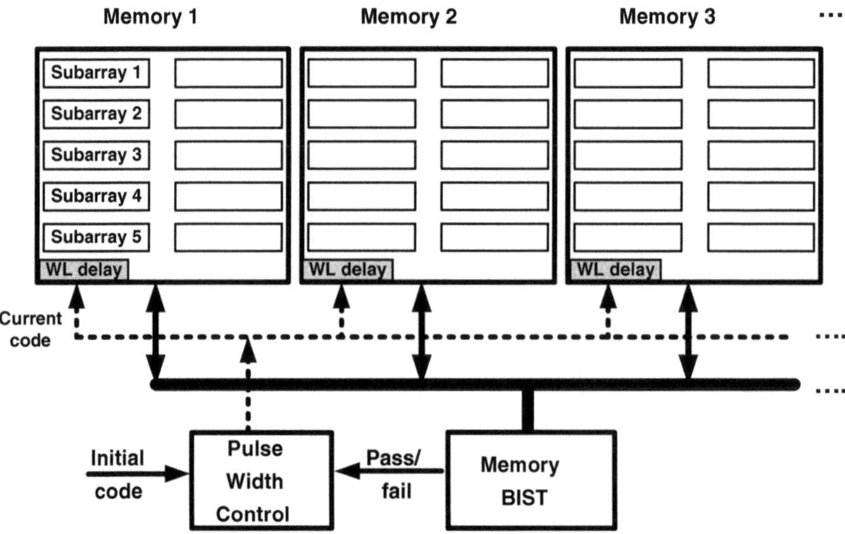

Fig. 4.6 Proposed architecture: Fine-grained wordline pulse width control

A post-silicon approach is required to reduce switching power due to the worst-case design practices used in SRAM design. Figure 4.6 shows a conceptual view of the proposed architecture. The system is composed of three main components: BIST, "WL delay", and "pulse width control" logic. The memory BIST is used to test the memory functionality by applying different testing patterns. Each memory instance contains a "WL delay" block, a digitally programmable delay element that controls the wordline pulse width T_{wl}. This adjustment for T_{wl} is achieved by adding the programmable delay element in the disable path of the wordline. The delay element is controlled using digital code provided by the "pulse width control" logic [35, 36]. The pulse width control logic varies the digital code depending on the BIST result. Therefore, the proposed system creates a closed feedback loop between the BIST, WL pulse width control, and the memory internal timing, which can be used to reduce the power consumption as explained below.

Next, we present more details on the three main components of the proposed system: BIST, "WL delay", and "pulse width control" blocks.

4.5.1 SRAM Built-in Self-Test

SRAMs are prone to different types of failure (catastrophic and parametric). To address yield problems in SRAMs, large arrays are designed with redundant rows and/or columns, which can be used to replace defected bitcells to repair the memory. To enable the repair, the memory should be tested first to locate the defective bitcells.

However, since embedded memories lack direct access to chip input/output signal, embedded memory testing becomes complicated and time consuming [37].

Memory BIST tests the memory using data patterns such as checkerboard, solid, row stripe, column stripe, double row stripe, and double column stripe. Different test algorithms are used to cover different types of faults; however, one of the most widely used algorithms is the MARCH test [34]. BIST not only reduces testing time and cost, but also allows testing the memory under actual clock speed which is known as at-speed testing. Memory self-test can be performed every time the chip is reset (in power-up test mode) [12, 34].

4.5.2 WL Programmable Delay Elements

Wordline and sense amplifier timing are of utmost importance for correct read operation. The WL programmable delay (also known as timer) shown in Fig. 4.3 is responsible for generating these critical internal timing signals. For SRAM post-silicon debugging and yield learning, programmable delay elements are used to control internal timing for wordline and sense amplifier [35, 36, 38, 39]. For example, these delay elements are used to characterize the margin in bitline differential voltage. Moreover, they can be used in relaxing address setup time requirements by delaying the clock rising edge [35]. Figure 4.7 shows one type of programmable delay elements used in timer circuits [35, 38].

4.5.3 Pulse Width Control Logic

The proposed architecture employs pulse width control logic, which can be as simple as a digital counter or a more complex state machine that varies the digital code for the delay element depending on the output of the BIST, or it can be implemented in software. For example, a programmable processor can be used to control both the BIST and the programmable delay elements. Fortunately, modern SoCs include programmable processors that can be used at start-up time to test and verify the operation of other modules [37, 40].

4.5.4 System Operation

Figure 4.8 shows the operation of the proposed system: Initially, the pulse width control logic provides the initial code for the memory. This initial digital code will correspond to the required worst case $T_{\mathrm{wl,wc}}$ for a given yield requirement, which is determined during the design phase (using Eq. (4.5)). The BIST tests the memory

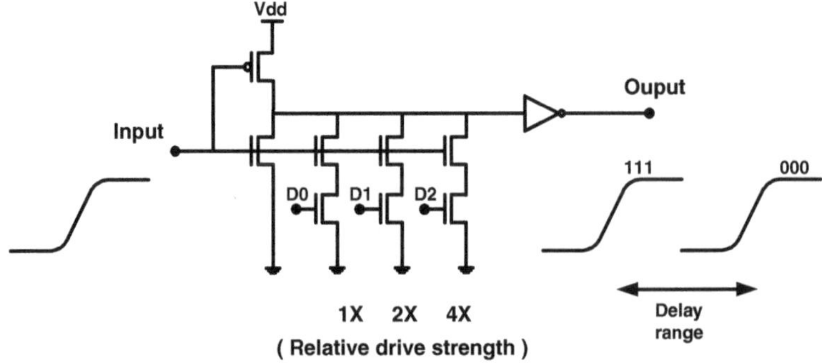

Fig. 4.7 Example of a programmable delay element [35]

using the initial code. If the memory fails, it may require repair or it may be discarded, which is a typical result of a BIST testing sequence. However, if the memory passes the BIST testing using the initial code (or after repair), the BIST signals the pulse width control logic to reduce the digital code, hence reducing T_{wl} using the programmable delay element. Using the new digital code for T_{wl}, the BIST tests the memory once again, and this process of BIST and T_{wl} reduction is repeated until the memory fails in read operation. The last passing code is then stored on the built-in registers inside the memory. If the code corresponding to the shortest T_{wl} is reached without the memory failing, the code is stored and the operation is terminated. The above mentioned steps are repeated for all the memories in the chip. Hence, the proposed architecture reduces T_{wl} for all memories in which the bitcells have sufficient ΔV_{bl} for correct read operation. Therefore, by reducing T_{wl}, the switching power of the memory can be reduced. This operation can be part of the system testing or power up, where the final codes for each memory can be stored on built-in registers or stored on fuses.

4.6 Results and Discussion

To estimate the power savings using the proposed architecture, Monte Carlo simulations were used to capture the impact of device variation on bitcell I_c and the corresponding T_{wl}. Simulations were performed using a 1 Mb macro from an industrial 45 nm technology. The 1 Mb macro uses a replica path to reduce power consumption and improve process tracking [15, 20–22]. Hardware correlated bitcell statistical models[1] are used to compute μ_{I_c} and σ_{I_c} which are used in the simulation flow

[1] The statistical models are provided by the foundry, and the statistical data are extracted from measuring a large sample of bitcells as shown in [27]. Random WID variations are reflected in the Spice model by varying V_{th}, W, L and other parameters of the bitcell devices according to the measured statistics.

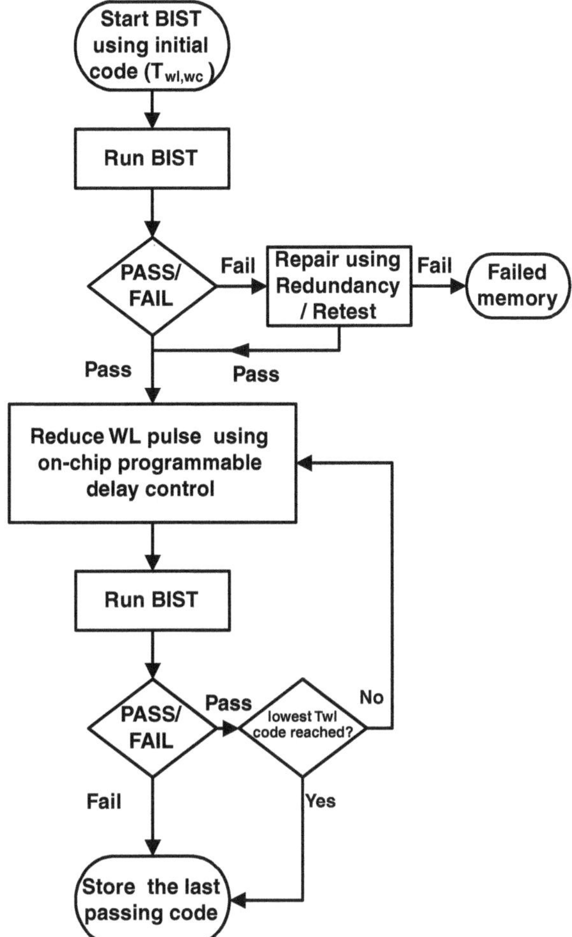

Fig. 4.8 Flowchart for the operation of the proposed fine-grained wordline pulse width control system

shown below. Post-layout switching power simulations were used to measure the power versus T_{wl} dependence, as shown in Fig. 4.2.

In SoC design, typically a high density macro (on the order of 512 k or 1 Mb) is used as a building block for larger memories. Therefore, in our simulations, we assume that the minimum memory macro size is 1 Mb, and multiple instances of that macro are used to realize larger memories.

To estimate power saving using the proposed system, a Monte Carlo simulation flow was developed as follows:

1. For every memory instance in the chip, generate N_{bit} samples of I_c normal distribution with mean μ_{I_c} and standard deviation σ_{I_c} to represent the read current variation in the macro.

2. Find the lowest bitcell current I_c in each memory instance and use it to compute the $T_{\mathrm{wl,inst}}$ using Eq. (4.3). Therefore, $T_{\mathrm{wl,inst}}$ represents the minimum wordline pulse width that guarantees correct read operation for *that instance*. This value of $T_{\mathrm{wl,inst}}$ should be automatically determined by the proposed system, since the wordline control block shown in Fig. 4.6 will reduce $T_{\mathrm{wl,inst}}$ until that memory instance fails a read operation.
3. Using Eq. (4.4), calculate the power consumption of that memory instance.
4. Repeat all the above steps for all memory instances on that chip, and add them to estimate the total read power for memories on that chip.[2]
5. Repeat all the above steps for a large number of chips, and determine the average power.
6. From the chip level yield requirement and the total memory density, calculate the target yield in terms of σ using Eq. (4.6) which is also used to find the worst case $T_{\mathrm{wl,wc}}$ using Eq. (4.5). This will be the value of T_{wl} that would have been used for all memory instances if the proposed architecture was not enabled. For $T_{\mathrm{wl,wc}}$, the corresponding power consumption P_{wc} can be calculated using Eq. (4.4).
7. From the last two steps, calculate the power reduction using the proposed architecture.

Figure 4.9 shows the power reduction achieved using the proposed architecture for memory densities and different target yields. Note that, these yield values represent the intrinsic yield before applying any repair or correction (i.e., redundancy or ECC). As the memory density increases, the power saving increases, which shows the effectiveness of the proposed system, especially if high yield is required (as in high volume, low cost products). Array switching power consumption can be reduced between 15 and 35 % for a 48 Mb memory density, depending on the required yield. Figure 4.10 shows the achievable power saving versus memory yield target. Power saving increases with memory yield; even for a 1 Mb memory density, the array switching power savings can be as high as 15 % for a yield of 99 %. As the SRAM content increases in future SoCs, it is expected that even greater power saving can be achieved using the proposed architecture.

SRAM bitcells are sensitive to device variations, which increase significantly with scaling. To study how higher variations affect power savings using the proposed system, we increased I_c variations from 10 to 12.5 % [41]. Even with increased I_c variations, the proposed system achieves significant power savings as shown in Fig. 4.11. For the 48 Mb case, power saving increases by more than two times, from 25 to 55 % (even though I_c variations only increased from 10 to 12.5 %). This shows the effectiveness of the proposed architecture in the presence of large process variations, which are expected to worsen with technology scaling.

[2] Here, we assume that all memories are accessed simultaneously. For our analysis, this is a fair assumption since we do not make any assumptions related to how the system accesses these memories. Nevertheless, the same simulation flow can be used if the switching activity for each memory is known.

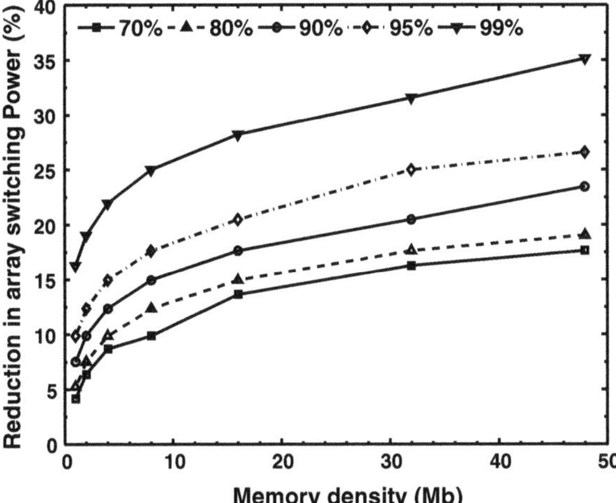

Fig. 4.9 Power reduction versus chip level memory density using the proposed architecture. Different values of memory target yield are shown

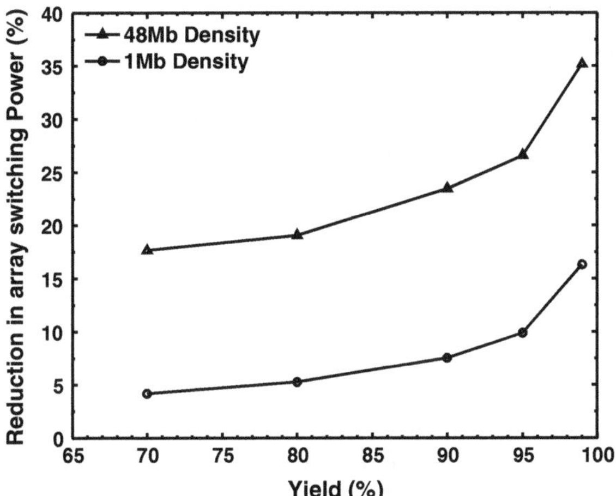

Fig. 4.10 Power reduction using the proposed architecture versus target yield for two cases of chip level density: **a** 1 Mb and **b** 48 Mb

For the previous simulation results, we assumed that a 1 Mb memory is the minimum memory instance size that can have a specific T_{wl}. However, in memory design, multiple subarrays (i.e., banks) are used to implement a single memory instance, as shown in Fig. 4.6. Typically, a timer circuit is used in each subarray, so the fine-grained concept can be further applied to the subarray level. This requires adding the digitally controlled delay element and the built-in storage registers for each subarray.

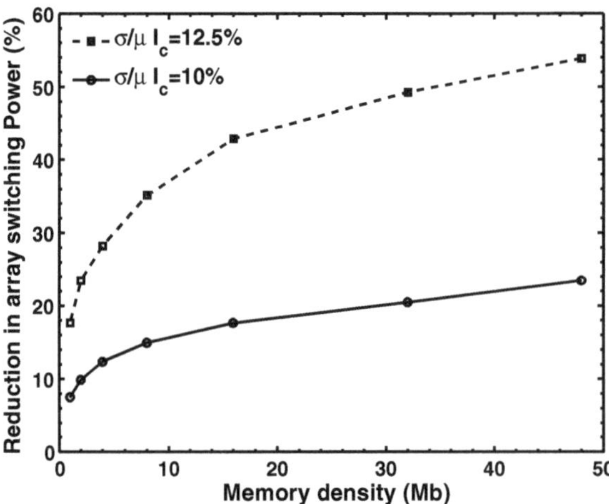

Fig. 4.11 Power reduction versus chip level memory density using the proposed architecture for different values of I_c variation for a target yield of 95 %

Nevertheless, the area overhead can still be very small since the additional area is distributed over the large area of the memory macro.

To evaluate the benefits of T_{wl} control at the subarray level, we assume that we have 16 subarrays in the 1 Mb macro. Each subarray is composed of 256 bitlines by 256 rows. Therefore, using the proposed architecture, T_{wl} can be adjusted for a 64 kb block. Figure 4.12 shows the achieved power saving for the 64 kb block as well as the 1 Mb full macro. By adjusting T_{wl} at the subarray level, power saving increases by 1.75X, from 24 to 42 %, for the same 48 Mb chip level memory density. Also for the 1 Mb chip level memory density, power saving increases from 7.5 to 20 % (2.67X improvement). This shows the importance of reducing the size of the memory block which can be individually controlled using T_{wl}, as this will significantly reduce the array switching power consumption.

Reducing T_{wl} affects hold time requirements since smaller T_{wl} reduces the access time of the memory T_{cq} which implies that the delay through the memory is now faster. Lower access time can cause hold violations in the flops in the output path of the memory. This situation can be prevented in the design phase by using the lowest expected T_{cq} for hold time verification (i.e., using the minimum T_{wl} that the programmable delay element can provide).

In the above analysis, the discrete quantization effect on power savings is not considered. In reality, since we are using a digitally controlled delay element, T_{wl} can only take discrete values. However, this may not have a significant impact on the power savings, since small area delay elements can cover large range of delays with fine control [35, 36]. In addition, by using the Monte Carlo simulation flow described above, we can determine the most probable range of T_{wl}, and modify the delay elements in design phase to have enough steps in that region. It is important to

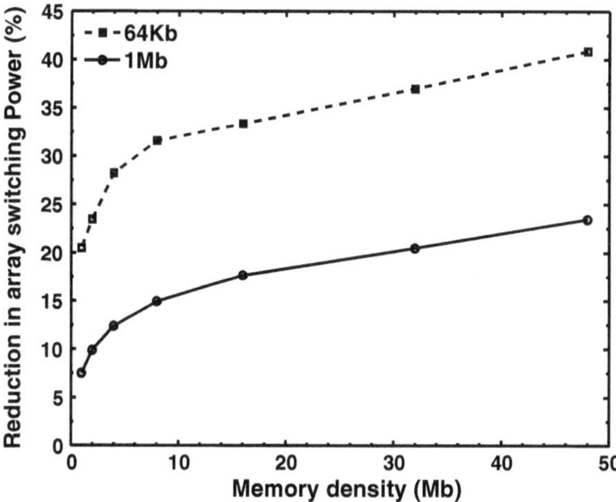

Fig. 4.12 Power reduction versus chip level memory density using the proposed architecture for different values of the minimum controlled memory instance (or subbanks) for a target yield of 95 %

note that delay elements do not add extra area, as they are typically used in memories for debugging purposes [35, 39, 36].

While the proposed system accounts for static process variations, it cannot adapt to dynamic environmental variations such as voltage or temperature variation [24] due to their dynamic nature. This results from the fact that T_{wl} will be fixed after running the pulse width control system in the start-up. Hence, environmental variations may cause the memory to fail if variations reduce the sensing margin. This problem can be addressed in design phase, by ensuring that the minimum step size for delay control provides sufficient margin for voltage and temperature variations. Moreover, if self-timed memories are used, the replica path automatically provides efficient tracking for environmental variations [15, 20–22]. In addition, in product testing, low voltage screening can be used in power-up to set T_{wl}, which guarantees a sufficient margin for environmental variations, since the product will operate at a supply voltage typically larger than the low voltage test condition.

In addition to environmental variations, transistor aging mechanisms such as hot carrier injection (HCI), time-dependent dielectric breakdown (TDDB), and negative bias temperature instability (NBTI) cause device characteristics to shift with time, which increases gate delay [42, 43]. To prevent the system from failing due to device aging, additional guard-banding (in delay or in the minimum supply voltage) is required, which may lower the power savings of the proposed architecture. For an accurate estimate of power savings including device aging, device reliability models for these effects need to be included in the proposed simulation flow.

In this work, we presented the analysis and results only for read power reduction. Nevertheless, the proposed system can be extended to reduce switching power in write operation. In write operation, one side of the bitline is pulled down to zero, while the other side is held at V_{dd} using write drivers. This operation is enabled for columns selected for write using a column multiplexer. However, half-selected bitcells (bitcells on the same wordline which are not accessed for write operation) still experience a dummy read operation, and the bitline discharges using I_c for a period determined by T_{wl}. Array switching power for write operation $P_{s_{array},w}$ can therefore be calculated as:

$$P_{s_{array},w} \simeq \begin{cases} [N_{IO}\, V_{dd} + (N_{bl} - N_{IO})\; I_c\, T_{wl}] \times N_{wl} C_{bit} V_{dd}\, f & \text{for } T_{wl} \leq \frac{V_{dd}\, N_{wl}\, C_{bit}}{I_c} \\ N_{bl}\, N_{wl}\, C_{bit}\, V_{dd}^{\,2}\, f & \text{for } T_{wl} > \frac{V_{dd}\, N_{wl}\, C_{bit}}{I_c} \end{cases} \tag{4.13}$$

where N_{IO} is the input/output data width for the memory.

The term including $N_{IO}\, V_{dd}$ in Eq. (4.13) refers to the power consumption in the selected bitlines for write operation, which are pulled down to zero using write drivers (not the bitcell read current I_c, so this term is not dependent on T_{wl}). The term including $(N_{bl} - N_{IO})$ is the power consumption in the half-selected bitlines in write operation, which is dependent on T_{wl}. Array switching power due to half-selected bitcells can contribute significantly to the total write power, especially in high density memories with small IO width and large muxing option (i.e., $N_{bl} \gg N_{IO}$). Using the same approach as in read operation, the proposed architecture can be used to reduce T_{wl} in write operation and reduce the power consumption of half-selected bitcells. However, as T_{wl} is reduced, the bitcell becomes more susceptible to write failures, which happens when the cell fails to write the desired value during write operation [26, 16]. Therefore, the proposed system can reduce T_{wl} until a write failure occurs. To evaluate the power savings in this case, a statistical simulation flow that includes models for write failure is required. To implement both read and write power reduction, additional logic is required to control T_{wl} for read and write separately as was shown in [39]. Nevertheless, the area penalty is negligible due to the small size of digital delay elements [35, 39, 36].

In addition to reducing power, reducing T_{wl} reduces read disturb failures [44–47]. Read disturb occurs when the bitcell flips its stored data when it accessed (or half-selected). As T_{wl} becomes smaller, the bitcell has less time to flip, which reduces the read disturb failure probability [44–47].

4.7 Summary

With the large increase in process variation, array switching power in future SRAMs are expected to increase to ensure correct read operation and achieve high yield. In this work, we propose a new memory architecture that significantly reduces array switching power while insuring high robustness against variations. The proposed architecture combines BIST and digitally controlled delay elements to reduce the wordline pulse width while ensuring correct read operation which reduces switching power. Combining both architecture and circuit techniques enables the system to detect weak bitcells using the BIST, and adjust T_{wl} accordingly. Therefore, the proposed architecture recovers the lost power consumption since it adjusts T_{wl} to ensure correct read operation for each memory block. Monte Carlo simulations using a 1 Mb SRAM macro from an industrial 45 nm technology were used to examine the power reduction achieved by the system. The proposed architecture reduces the array switching power significantly, especially as the chip level memory density increases. For a 48Mb memory density, the architecture yields a 27 % reduction in array switching power for a target yield of 95 %. The proposed system can provide larger power saving as process variations increase, which makes it an attractive option for future technologies.

References

1. K. Itoh, M. Horiguchi, M. Yamaoka, Low-voltage limitations of memory-rich nano-scale CMOS LSIs, in *37th European Solid State Device Research Conference (ESSDERC 2007)*, Sept 2007, pp. 68–75
2. A. Agarwal, B. Paul, H. Mahmoodi, A. Datta, K. Roy, A process-tolerant cache architecture for improved yield in nanoscale technologies. IEEE Trans. Very Large Scale Integr. VLSI Syst. **13**(1), 27–38
3. Y.-C. Lai, S.-Y. Huang, H.-J. Hsu, Resilient self-v -tuning scheme with speed-margining for low-power SRAM. IEEE J. Solid-State Circuits **44**(10), 2817–2823
4. S. Mukhopadhyay, K. Kim, H. Mahmoodi, K. Roy, Design of a process variation tolerant self-repairing SRAM for yield enhancement in nanoscaled CMOS. IEEE J. Solid-State Circuits **42**(6), 1370–1382
5. M. Yamaoka, N. Maeda, Y. Shimazaki, K. Osada, 65nm low-power high-density SRAM operable at 1.0 v under 3σ; systematic variation using separate Vth monitoring and body bias for NMOS and PMOS", in *IEEE international Solid-State Circuits Conference (ISSCC 2008)*, Digest of Technical Papers, Feb 2008, pp. 384–622
6. O. Hirabayashi, A. Kawasumi, A. Suzuki, Y. Takeyama, K. Kushida, T. Sasaki, A. Katayama, G. Fukano, Y. Fujimura, T. Nakazato, Y. Shizuki, N. Kushiyama, T. Yabe, A process-variation-tolerant dual-power-supply SRAM with 0.179 um2 cell in 40 nm CMOS using level-programmable wordline driver", in *IEEE international Solid-State Circuits Conference—Digest of Technical Papers (ISSCC 2009)*, Feb 2009, pp. 458–459,459a
7. E. Karl, Y. Wang, Y.-G. Ng, Z. Guo, F. Hamzaoglu, U. Bhattacharya, K. Zhang, K. Mistry, M. Bohr, A 4.6 GHz 162 Mb SRAM design in 22 nm tri-gate CMOS technology with integrated active VMIN-enhancing assist circuitry. in *IEEE International Solid-State Circuits Conference Digest of Technical Papers (ISSCC)*, Feb 2012, pp. 230–232
8. M. Abu-Rahma, M. Anis, and S. S. Yoon, Reducing SRAM power using fine-grained wordline pulsewidth control. IEEE Trans. Very Large Scale Integr. VLSI Syst. **18**(3), 356–364
9. F. Kurdahi, A. Eltawil, Y.-H. Park, R. Kanj, S. Nassif, System-level SRAM yield enhancement. in *7th International Symposium on Quality Electronic Design (ISQED '06)*, March 2006, pp. 6–184
10. I. J. Chang, D. Mohapatra, K. Roy, A voltage-scalable and process variation resilient hybrid SRAM architecture for MPEG-4 video processors, in *46th ACM/IEEE Design Automation Conference (DAC '09)*, July 2009, pp. 670–675
11. A. Agarwal, B. Paul, S. Mukhopadhyay, K. Roy, Process variation in embedded memories: failure analysis and variation aware architecture. IEEE J. Solid-State Circuits **40**(9), 1804–1814

12. Y. Zorian, S. Shoukourian, Embedded-memory test and repair: infrastructure IP for SoC yield. IEEE Des. Test Comput. **20**(3), 58–66 (2003)
13. M. Pelgrom, H. Tuinhout, M. Vertregt, Transistor matching in analog CMOS applications. in *Proceedings of the International Electron Devices Meeting (IEDM) 1998*, pp. 915–918
14. T.-C. Chen, Where is CMOS going: trendy hype versus real technology. in *Proceedings of the International Solid-State Circuits Conference (ISSCC) 2006*, pp. 22–28
15. R. Heald, P. Wang, Variability in sub-100 nm SRAM designs. in *Proceedings of International Conference on Computer Aided Design 2004*, pp. 347–352
16. K. Agarwal, S. Nassif, Statistical analysis of SRAM cell stability. in *Proceedings of the 43rd Annual Conference on Design automation 2006 (DAC '06)*, pp. 57–62
17. R. Venkatraman, R. Castagnetti, S. Ramesh, The statistics of device variations and its impact on SRAM bitcell performance, leakage and stability", in *Proceedings of the International Symposium on Quality of Electronic Design (ISQED)*, 2006, pp. 190–195
18. M. Q. Do, M. Drazdziulis, P. Larsson-Edefors, L. Bengtsson, Parameterizable architecture-level SRAM power model using circuit-simulation backend for leakage calibration, in *Proceedings of the International Symposium on Quality of Electronic Design (ISQED)*, 2006, pp. 557–563
19. A. Macii, L. Benini, M. Poncino, *Memory Design Techniques for Low Energy Embedded Systems* (Kluwer Academic Pub, 2002)
20. B. Amrutur, M. Horowitz, A replica technique for wordline and sense control in low-power SRAM's. IEEE J. Solid-State Circuits **33**(8), 1208–1219 (1998)
21. K. Osada, J.-U. Shin, M. Khan, Y.-D. Liou, K. Wang, K. Shoji, K. Kuroda, S. Ikeda, K. Ishibashi, Universal-Vdd 0.65–2.0 V 32 kB cache using voltage-adapted timing-generation scheme and a lithographical-symmetric cell. in *Proceedings of the International Solid-State Circuits Conference (ISSCC 2001)*, pp. 168–169, 443
22. M. Yamaoka, N. Maeda, Y. Shinozaki, Y. Shimazaki, K. Nii, S. Shimada, K. Yanagisawa, T. Kawahara, Low-power embedded SRAM modules with expanded margins for writing. in *Proceedings of the International Solid-State Circuits Conference (ISSCC 2005)*, Vol 1, pp. 480–611
23. D. Marculescu, E. Talpes, Variability and energy awareness: a microarchitecture-level perspective. in *Proceedings of the 42nd Annual Conference on Design automation (DAC '05)*, 2005, pp. 11–16
24. S. Borkar, T. Karnik, S. Narendra, J. Tschanz, A. Keshavarzi, V. De, Parameter variations and impact on circuits and microarchitecture. in *Proceedings of the 40th conference on Design automation (DAC '03)*, 2003, pp. 338–342
25. M. H. Abu-Rahma, K. Chowdhury, J. Wang, Z. Chen, S. S. Yoon, M. Anis, A methodology for statistical estimation of read access yield in SRAMs. in *Proceedings of the 45th Conference on Design Automation (DAC '08)*, 2008, pp. 205–210
26. S. Mukhopadhyay, H. Mahmoodi, K. Roy, Statistical design and optimization of SRAM cell for yield enhancement. in *Proceedings of International Conference on Computer Aided Design (DAC '08)*, 2004, pp. 10–13
27. T. Fischer, C. Otte, D. Schmitt-Landsiedel, E. Amirante, A. Olbrich, P. Huber, M. Ostermayr, T. Nirschl, J. Einfeld, A 1 Mbit SRAM test structure to analyze local mismatch beyond 5 sigma variation. in *Proceedings of the IEEE International Conference on Microelectronic Test Structures (ICMTS '07)*, 2007, pp. 63–66
28. R. Rao, A. Srivastava, D. Blaauw, D. Sylvester, Statistical analysis of subthreshold leakage current for VLSI circuits. IEEE Trans Very Large Scale Integr. Syst. **12**(2), 131–139 (2004)
29. J. Wang, P. Liu, Y. Gao, P. Deshmukh, S. Yang, Y. Chen, W. Sy, L. Ge, E. Terzioglu, M. Abu-Rahma, M. Garg, S. S. Yoon, M. Han, M. Sani, G. Yeap, Non-Gaussian distribution of SRAM read current and design impact to low power memory using voltage acceleration method, in *Symposium on VLSI Technology (VLSIT)*, June 2011, pp. 220–221
30. P. Kinget, Device mismatch and tradeoffs in the design of analog circuits. IEEE J. Solid-State Circuits **40**(6), 1212–1224 (2005)
31. B. Wicht, T. Nirschl, D. Schmitt-Landsiedel, Yield and speed optimization of a latch-type voltage sense amplifier. IEEE J. Solid-State Circuits **39**(7), 1148–1158 (2004)

32. S. Mukhopadhyay, K. Kim, K. Jenkins, C.-T. Chuang, K. Roy, Statistical characterization and on-chip measurement methods for local random variability of a process using sense-amplifier-based test structure. in *Proceedings of the International Solid-State Circuits Conference (ISSCC)*, Feb 2007, pp. 400–611
33. A. Papoulis, *Probability, Random Variables, and Stochastic Processes*, 3rd edn. (McGraw-Hill, 1991)
34. R. Rajsuman, Design and test of large embedded memories: an overview. IEEE Des. Test Comput. **18**(3), 16–27 (2001)
35. W. Kever, S. Ziai, M. Hill, D. Weiss, B. Stackhouse, A 200 MHz RISC microprocessor with 128 kB on-chip caches. in *Proceedings of the International Solid-State Circuits Conference (ISSCC)*, 6–8 Feb 1997, pp. 410–411,495
36. Y. H. Chan, T. J. Charest, J. R. Rawlins, A. D. Tuminaro, J. K. Wadhwa, O. M. Wagner, Programmable sense amplifier timing generator. U. S. Patent 6,958,943, Oct 2005
37. A. Chandrakasan, W.J. Bowhill, F. Fox, *Design of High-Performance Microprocessor Circuits* (Wiley-IEEE Press, 2000)
38. M. Min, P. Maurine, M. Bastian, M. Robert, A novel dummy bitline driver for read margin improvement in an eSRAM. in *Proceedings of the IEEE International Symposium on Electronic Design, Test and Applications DELTA*, Jan 2008, pp. 107–110
39. R. Joshi, R. Houle, D. Rodko, P. Patel, W. Huott, R. Franch, Y. Chan, D. Plass, S. Wilson, S. Wu, R. Kanj, A high performance 2.4 Mb L1 and L2 cache compatible 45nm SRAM with yield improvement capabilities. in *Proceedings of IEEE Symposium on VLSI Circuits*, June 2008, pp. 208–209
40. J. M. Rabaey, A. Chandrakasan, B. Nikolic, *Digital Integrated Circuits*, 2nd Edn. (Prentice Hall, 2002)
41. H. Yamauchi, Embedded SRAM circuit design technologies for a 45 nm and beyond. in *7th International Conference on ASIC (ASICON '07)*, 22–25 Oct 2007, pp. 1028–1033
42. K. Kang, S.P. Park, K. Roy, M.A. Alam, Estimation of statistical variation in temporal NBTI degradation and its impact on lifetime circuit performance. in *Proceedings of the IEEE/ACM international Conference on Computer-aided design (ICCAD '07)*, 2007, 730–734
43. X. Li, J. Qin, B. Huang, X. Zhang, J. Bernstein, SRAM circuit-failure modeling and reliability simulation with SPICE. IEEE Trans. Device Mater. Reliab. **6**(2), 235–246 (2006)
44. M. Khellah, Y. Ye, N. Kim, D. Somasekhar, G. Pandya, A. Farhang, K. Zhang, C. Webb, V. De, Wordline and bitline pulsing schemes for improving SRAM cell stability in low Vcc 65 nm CMOS designs, in *Proceedings of IEEE Symposium on VLSI Circuits*, 2006, pp. 9–10
45. S. Ikeda, Y. Yoshida, K. Ishibashi, Y. Mitsui, Failure analysis of 6 T SRAM on low-voltage and high-frequency operation. IEEE Trans. Electron Devices **50**, 1270–1276 (2003)
46. J. B. Khare, A. B. Shah, A. Raman, G. Rayas, Embedded memory field returns—trials and tribulations. in *IEEE International Test Conference (ITC '06)*, Oct 2006, pp. 1–6
47. M. Yamaoka, K. Osada, T. Kawahara, A cell-activation-time controlled SRAM for low-voltage operation in DVFS SoCs using dynamic stability analysis. in *34th European Solid State Circuits Conference (ESSCIRC)*, Sept 2008, pp. 286–289

Chapter 5
A Methodology for Statistical Estimation of Read Access Yield in SRAMs

5.1 Challenges of SRAM Statistical Design

Random variations in nanometer range technologies are considered one of the most important design considerations [1, 2]. This is especially true for SRAM memories, due to the large variations in bitcell characteristics. Typically, SRAM bitcells utilize the smallest device sizes on a chip. Therefore, they show the largest sensitivity to different sources of variations—such as random dopant fluctuations (RDF), line-edge roughness (LER), and others [3, 4]. While variations in logic circuits have been shown to cause delay spread [5, 6], which reduces parametric yield, for SRAMs, process variations also cause the memory to functionally fail, which reduces the chip's functional yield. With lower supply voltages and higher variations, statistical simulation methodologies become imperative to estimate memory yield and optimize performance and power.

As explained in Sect. 2.6, there are different types of SRAM bitcell failure mechanisms, such as static noise margin stability fails (cell may flip when accessed), write fails (bitcell cannot be written within the write window), read access fails (incorrect read operation), and retention fails [4, 7, 8]. Statistical simulations are used to estimate the probability of failure, or inversely yield, for different failure mechanisms. To achieve high yield, the bitcell failure probability must be very low ($10^{-9} - 10^{-12}$) or $5.5 - 6.5\sigma$; however, due to the large density of embedded memories, this small failure probability can have a large impact on yield. From a simulation point of view, it is extremely challenging to estimate probabilities in this small range since it requires a huge number of simulation runs which are computationally intensive, and in most cases impractical. In the following sections, we review statistical simulation approaches used in SRAM design to address this problem.

M. H. Abu-Rahma and M. Anis, *Nanometer Variation-Tolerant SRAM*, DOI: 10.1007/978-1-4614-1749-1_5,

5.2 Estimating SRAM Failure Probability

Algorithms for SRAM yield estimation typically employ three steps [9–11]:

- Generation of random samples that follow the process distribution. This step is typically not computationally intensive.
- Evaluating or simulating the impact of variations on the circuit. For example, SPICE simulation or compact models can be used to estimate the impact of variations. This step is usually time-consuming since it involves simulating the impact of variation on the circuit.
- Calculate yield depending on the failure criteria (i.e., find the number of samples that meet the specification) from step 2.

Methods to speed up statistical simulation are intended to reduce the evaluation time (step 2) by either reducing the number of required samples or using a simplified model for evaluation. In the following sections we explore both techniques.

5.2.1 Direct Monte Carlo

Standard Monte Carlo is one of the most commonly used techniques in statistical yield estimation. Assume we are interested in estimating the yield of an SRAM where a performance metric $f(x)$ is defined (e.g., SNM, WM, I_{read}, etc.), and x is the random variable (e.g., V_{th}).[1] Also, assume that failure criteria is defined as $f(x) > f_c$, where f_c is the minimum specification that the circuit should meet for correct operation. Therefore, the indicator function $I(x)$ (pass/fail function) is defined as [11]:

$$I(x) = \begin{cases} 0, & \text{pass} \quad (f(x) < f_c) \\ 1, & \text{fail} \quad (f(x) > f_c) \end{cases} \tag{5.1}$$

and the failure probability P_f can be defined as:

$$P_f = P(f(x) > f_c) = \int_{-\infty}^{\infty} I(x)\text{pdf}_x(x)\text{d}x \tag{5.2}$$

where, $\text{pdf}_x(x)$ is the probability density function of x, the parameter which is varying.

The challenge in predicting P_f is due to the fact that $I(x)$ is unknown since the boundary between pass and fail is difficult to find, especially when there are multiple variables (i.e., multidimensional space), which is usually the case.

---

[1] Here, we assume that there is only random variable that affects circuit operation, however, the concept can be easily extended to multiple dimensions.

The Monte Carlo approach is generally used to evaluate the integral shown in Eq. (5.2), as follows:

$$P_f = P(f(x) > f_c) = \int_{-\infty}^{\infty} I(x)\mathrm{pdf}_x(x)\mathrm{d}x \approx \frac{1}{N}\sum_{j=1}^{N} I(x_j) \tag{5.3}$$

In this approach, a large number of samples are generated that follow the distribution of the process variation parameters ($\mathrm{pdf}_x(x)$). Next, these samples are simulated to evaluate the required performance metric [9, 11]. The failure probability can be estimated by direct evaluation of the SRAM metrics using Monte Carlo, and then extrapolating to long tails by assuming a particular distribution (i.e., assuming SNM is a Gaussian distribution). The technique has been used to evaluate static read stability, writeability, and read access (sense margin) [7].

5.2.2 Errors Associated with Monte Carlo

As mentioned in the previous section, the failure probability using Monte Carlo simulation can be estimated as:

$$P_f = \frac{N_f}{N} \tag{5.4}$$

where N_f is the number of samples that do not meet the performance metric out of N total simulations. P_f is a Bernoulli distribution, therefore, the variance of P_f can be calculated as [9]:

$$\sigma_{P_f}^2 = \frac{P_f(1 - P_f)}{N} \tag{5.5}$$

Due to the errors associated with statistical sampling, confidence intervals are usually defined to put bounds on the statistical estimate. For large N and P_F ($N \times P_f \gg 1$) and using the Central Limit Theorem, for a given confidence interval ε, and acceptable error criteria d, the required number of Monte Carlo samples N is:

$$N = \frac{\left(\Phi^{-1}(1-\varepsilon)\right)^2}{d^2}\frac{1 - P_f}{P_f} \tag{5.6}$$

where $\Phi^{-1}()$ is the inverse of the standard normal distribution. Assuming that a 95 % confidence is required, and the acceptable error criteria is 5 %, therefore, for a small P_f, Eq. (5.7) becomes:

$$N \approx \frac{384}{P_f} \tag{5.7}$$

Thus, the required number of samples N is inversely proportional to P_f. For practical ranges of P_f used in SRAM design, N becomes extremely large. Table 5.1 shows the required number of Monte Carlo for different values of P_f. The first column shows the probability of failure in terms of standard deviations for a normal distribution. In addition, the table shows estimated memory yield for different densities assuming that SRAM bitcell failure is the dominant failure mechanism (Yield $= (1 - P_f)^{N_{\text{cells}}}$). For example, for a 100 Mbit memory, if the target yield is >98 %, the failure probability should be lower than $4e^{-11}$, which corresponds to a 6.25σ design. To verify this very low P_f, more than $1e^{12}$ simulation runs are required, which is practically impossible. Therefore, different methods to reduce the number of required samples have been proposed. In the following sections, we review a few of these methods.

5.2.3 Compact Modeling

One technique to reduce simulation time is to use compact modeling. In this approach, a simplified model is developed to estimate SRAM metrics directly without circuit simulation. Since the compact model is much simpler than circuit simulation, a large number of Monte Carlo runs can be performed in a short simulation time. Physical models to estimate the impact of random V_{th} variations on SRAM static noise margin have been derived [12, 13]. Also, the compact modeling approach has been used to analyze subthreshold SRAM stability, where the voltage transfer characteristics (VTC) are modeled as a function of basic device parameters. Using graphical or numerical solutions, the SNM can be calculated [14, 15]. The compact modeling approach can provide a good estimate of the circuit's basic behavior and its dependence on key parameters. Although compact models substantially improve simulation efficiency, they can be inaccurate since many approximations are made to reach the simplified form of the compact model.

Similar to compact modeling, behavioral modeling has been proposed to speed up Monte Carlo simulation and predict read access yield [16]. More details about this approach will be discussed in Sect. 5.4. Performance models using response surface methodology (RSM) have also been used to facilitate probability extraction for SRAM parametric failures [17].

5.2.4 Sensitivity Analysis

Sensitivity analysis is a widely known technique to analyze the impact of variation on circuit performance. In this approach, the parameter of interest is approximated around the nominal point using multi-variable Taylor series, and is used to estimate the mean and variance of the performance metric [18]. For example, in the case of SNM and assuming that V_{th} due to local variations are independent, sensitivity can be used as follows [13, 19–21]:

Table 5.1 Required number of Monte Carlo runs required to estimate failure probability with a 95 % confidence and 10 % error calculated using Eq. (5.7)

$\Phi^{-1}(P_f)$	P_f	MC runs (N)	Memory yield						
			1 Kbit (%)	10 Kbit (%)	100 Kbit (%)	1 Mbit (%)	10 Mbit (%)	100 Mbit (%)	1 Gbit (%)
3	1.4×10^{-3}	2.84×10^{5}	25	0	0	0	0	0	0
3.25	5.8×10^{-4}	6.65×10^{5}	55	0	0	0	0	0	0
3.5	2.3×10^{-4}	1.65×10^{6}	79	9	0	0	0	0	0
3.75	8.8×10^{-5}	4.34×10^{6}	91	40	0	0	0	0	0
4	3.2×10^{-5}	1.21×10^{7}	97	72	4	0	0	0	0
4.25	1.1×10^{-5}	3.59×10^{7}	99	90	33	0	0	0	0
4.5	3.4×10^{-6}	1.13×10^{8}	100	97	71	3	0	0	0
4.75	1.0×10^{-6}	3.78×10^{8}	100	99	90	34	0	0	0
5	2.9×10^{-7}	1.34×10^{9}	100	100	97	74	5	0	0
5.25	7.6×10^{-8}	5.05×10^{9}	100	100	99	92	45	0	0
5.5	1.9×10^{-8}	2.02×10^{10}	100	100	100	98	82	14	0
5.75	4.5×10^{-9}	8.61×10^{10}	100	100	100	100	95	63	1
6	9.9×10^{-10}	3.89×10^{11}	100	100	100	100	99	90	35
6.25	2.1×10^{-10}	1.87×10^{12}	100	100	100	100	100	98	80
6.5	4.0×10^{-11}	9.57×10^{12}	100	100	100	100	100	100	96
6.75	7.4×10^{-12}	5.20×10^{13}	100	100	100	100	100	100	99
7	1.3×10^{-12}	3.00×10^{14}	100	100	100	100	100	100	100

$$\mu_{\text{SNM}} \approx \text{SNM}_0 + \frac{1}{2}\sum_{i=1}^{N}\left[\frac{\partial^2 \text{SNM}}{\partial {V_{\text{th},i}}^2}\right]\sigma^2_{V_{\text{th},i}} \tag{5.8}$$

$$\sigma^2_{\text{SNM}} \approx \sum_{i=1}^{N}\left[\frac{\partial \text{SNM}}{\partial V_{\text{th},i}}\right]^2 \sigma^2_{V_{\text{th},i}} \tag{5.9}$$

where μ_{SNM} and σ_{SNM} are the mean and standard deviation of SNM, respectively, SNM_0 is the nominal static noise margin (assuming no variations), $\partial\text{SNM}/\partial V_{\text{th},i}$ are the partial derivatives of SNM, which represent the sensitivity of SNM to V_{th} of individual transistors, N is the number of transistors in the circuit. Figure 5.1 shows the impact of V_{th} variations on SNM.

Sensitivity analysis simplifies the yield estimation problem considerably, since it reduces the number of simulations to only those necessary to estimate the sensitivities (partial derivatives) around the nominal point ($N + 1$ simulation runs, where N is the number of independent variables). However, the technique can introduce large inaccuracies since approximations using Taylor expansion is less accurate away from the nominal points.

Figure 5.1 shows the impact of V_{th} variations on SNM, which shows that the linearity assumptions of SNM is not accurate for large V_{th} variations. To overcome the limitations of conventional sensitivity analysis, modified sensitivity approaches have been proposed as discussed in [20]. Nevertheless, even with improved sensitivity analysis the accuracy may be limited. This inaccuracy results from the assumption that the impact of variations on the circuit is linearized around the nominal point, so the performance metric is assumed to follow the same distribution as the variation parameter (i.e., since V_{th} follows a Gaussian distribution, then SNM will also be Gaussian.) This approximation can introduce significant errors since the device operation is nonlinear, and becomes more nonlinear with lower supply voltage operation and higher variations.

5.2.5 Importance Sampling

In direct Monte Carlo simulation, many generated samples are near the mean of the distribution, which is usually not where failure occurs (typically near the tail of the distribution). Therefore, alternative methods have been proposed that allows better spread across the variation space such as Quasi Monte Carlo [9, 11, 22]. One of the widely used techniques to speed up Monte Carlo simulation is importance sampling (IS) which addresses some of the limitations of conventional Monte Carlo. In IS, the distributions of the variation parameters (e.g., V_{th} of a transistor) are intentionally distorted to produce more samples around the failure region. This distortion provides faster estimation of the failure probability with a smaller number of samples [9, 11, 22].

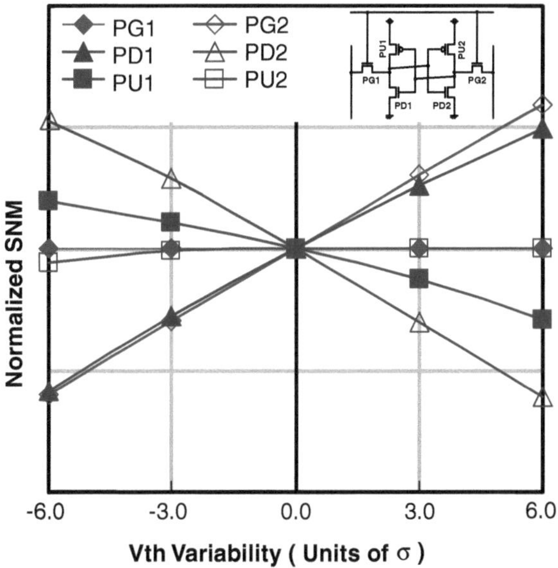

Fig. 5.1 Sensitivity analysis showing the impact of $V_{\rm th}$ variation on SNM [20]

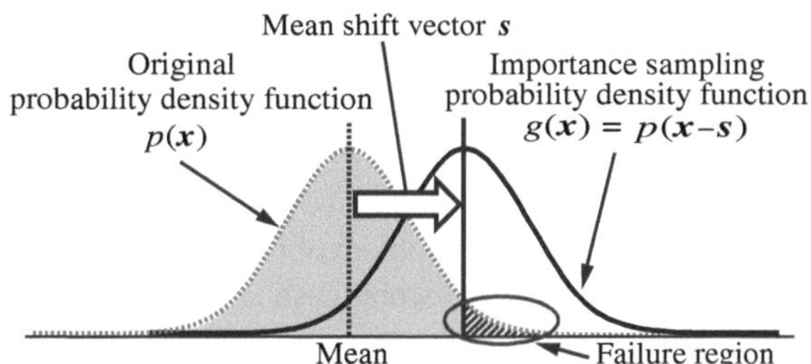

Fig. 5.2 Importance sampling using mean-shift approach [23]

Mathematically, estimating the failure probability using IS can be given by modifying Eq. (5.2) as follows [9, 11, 17, 22]:

$$P_f = P(f(x) > f_c) = \int_{-\infty}^{\infty} I(x)\frac{\mathrm{pdf}_x(x)}{g(x)}g(x)\mathrm{d}x \approx \frac{1}{N}\sum_{j=1}^{N}\left[I(x_j)\frac{\mathrm{pdf}_x(x_j)}{g(x_j)}\right] \tag{5.10}$$

where $g(x)$ is the distorting probability density function.

One disadvantage of IS is the ambiguity concerning how the distorting function, $g(x)$, should be chosen. For example, if $g(x)$ is chosen as a uniform distribution, the efficiency of the algorithm decreases as the dimensionality increases (i.e., increase in number of variation parameters) [22]. A commonly used distorted sampling function is derived by shifting the distribution to center around the failure region, by setting

$g(x) = \mathrm{pdf}_x(x - \mu_s)$, which is called a mean-shift IS approach [23], and μ_s is the applied shift. Figure 5.2 shows a representation of mean-shift IS.

The efficiency of IS algorithms relies on the choice of the distorted function that is used to generate random samples [24]. To improve yield prediction accuracy, the distorted PDF needs to be chosen so that most of the samples are located in the region where a failure is most likely to occur. Practically, this is difficult to achieve, since the failure region is difficult to find, especially when the problem involves a large number of parameters. To overcome this limitation, several techniques have been proposed. One technique involves using a mixture of distributions, where both uniform distribution ($U(x)$) and mean-shift IS are applied as follows [22]:

$$g_\lambda(x) = \lambda_1 \mathrm{pdf}_x(x) + \lambda_2 U(x) + (1 - \lambda_1 - \lambda_2)\mathrm{pdf}_x(x - \mu_s) \tag{5.11}$$

and $0 \leq \lambda_1 + \lambda_2 < 1$. λ_1 and λ_2 are chosen dependent on the location of μ_s [22].

Norm minimization has been used to reduce the variance of a mean-shift IS method [25]. Other improvements in IS involve using a pre-process technique used to find a robust shift-vector to be applied for mean-shift IS [23]. Other IS techniques use the Gibbs sampling method and utilize an optimization engine to find the distorted PDF using an adaptive sampling [24].

5.2.6 Most Probable Failure Point

Most probable failure point (MPFP) is another method used to analyze SRAM failure probability [26]. MPFP transforms the problem of finding the failure probability into an optimization problem that finds the worst-case vector of normalized variations that maximizes P_{fail} defined as:

$$p_{\mathrm{fail}} = \prod_{i=1}^{6} P(\Delta V_{\mathrm{th},i} \geq k_i \sigma_{V_{\mathrm{th},i}}) \tag{5.12}$$

where k_i denotes the normalized variation for the SRAM transistors' V_{th} with respect to its standard deviation at the MPFP. In other words, the k_i vector denotes the coordinates of the MFPF.

In this approach, the SRAM failure probability is estimated by finding a variation point that is most likely to cause the cell to fail. Figure 5.3 shows a 2D representation for an MPFP for two independent variation parameters. The X and Y axes show the normalized variation (e.g., V_{th} of pass gate and pull-down devices), which are assumed normally distributed, and with variation normalized to the standard deviation. The distance from the origin to the MPFP point represents the failure probability (in a number of standard deviations).

A simple way to find the MPFP is to divide the variation space into a large number of points and simulate at each point to locate the MPFP. This brute-force approach

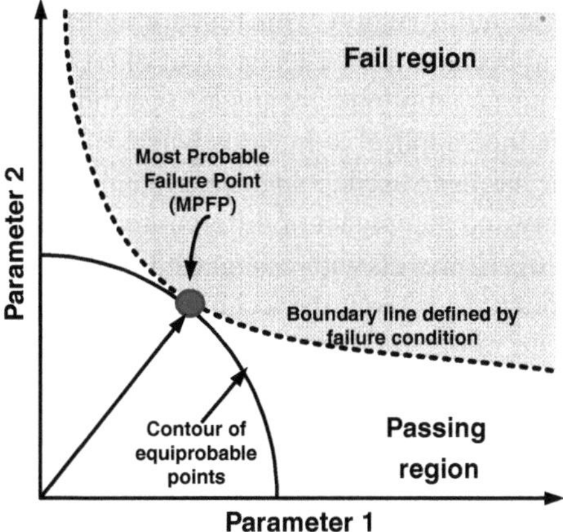

Fig. 5.3 Illustration of most-probable failure point (MPFP) in a 2D space [26]

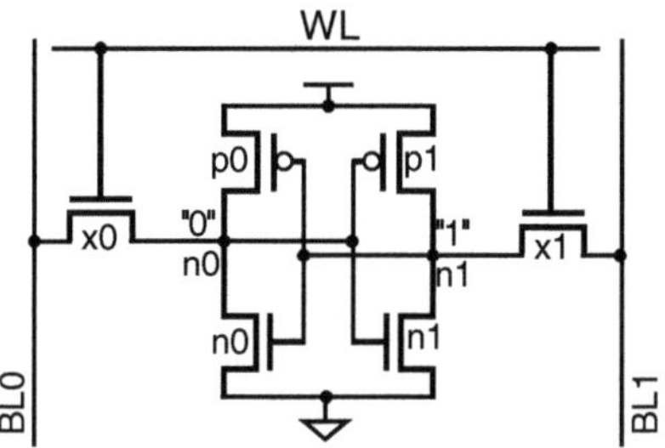

Failure Mechanism	Transistor V_{TH} variations					
	x0	n0	p0	x1	n1	p1
Read Disturb	-	+	-	+	-	+
Read Access	+	+	-	+	-	+
Write	+	-	+	+	+	-

Fig. 5.4 By considering failure sensitivity to device variation, the search space can be reduced by focusing only on the regions in which the failure is highly likely to occur [26]

becomes impractical for even a small number of variation parameters due to excessive computational time requirements.

In the case of SRAM 6T transistors, and assuming that local variations can be lumped into on variation parameter (V_{th}) per transistor, the variation space is then a 6-D space with orthogonal axes and 64 space regions. One way to improve the search approach is to limit the variation space by focusing only on the regions in which the failure is highly likely to occur. As shown in Fig. 5.4, for each failure mechanism, there is a most probable failure region. This reduction of variation space speeds up the searching algorithm because it focuses on 1 out of 64 space regions.

In this technique, no assumptions are made regarding the failure distribution, which makes this technique applicable in cases where the failure distribution is not known. This technique has been used to analyze dynamic stability and the impact of supply noise on stability and the impact of different read and write assist techniques on SRAM stability [26]. However, with a larger number of parameters, the search algorithm is more difficult. Moreover, the technique requires gradient and curvature information as part of the optimization step, which are not always available. Another limitation is that the optimizer may suffer from convergence especially when applied to objectives computed from nonlinear circuit simulation [27].

5.2.7 Statistical Blockade

Statistical blockade is a relatively new Monte Carlo-based method that uses a generalized approach to deal with rare events. The method employs ideas from extreme value theory (EVT) and machine learning to significantly improve the efficiency of Monte Carlo estimation [27].

EVT is a branch of probability that studies extreme or rare events, and has wide applications. Early work in the area of EVT demonstrated that the expected behavior of the maximum or minimum value of a sample of a distribution can be predicted using generalized extreme value distribution (GEV) [27]. In the special case of Gaussian samples, the EVT distribution is given by a Gumbel distribution, which has been applied to estimate SRAM read access yield in [28]. Statistical Blockade uses EVT to show that for a large class of distributions, we can make accurate predictions for the extrapolated tail of the distribution by fitting enough points in the tail using a generalized pareto distribution (GPD) [27].

Statistical blockade uses classifiers that can quickly be applied on a large number of samples to filter out the critical samples while blocking those that are unlikely to fall in the failure region, as shown in Fig. 5.5. Therefore, only a small set of Monte Carlo simulations are performed in the tail of the distribution of the performance metric of interest.

Statistical blockade filtering is accomplished as follows [27]:

1. Initial sampling generates data to build the classifier, as shown in Fig. 5.6. The initial sampling can use standard Monte Carlo or other sampling techniques.
2. Using the initial samples a classifier is built that uses a threshold point (specification) which is relaxed compared to the required threshold point.
3. More MC samples are generated but only those that are classified as fail points are simulated, as shown in Fig. 5.6.
4. The simulated data is then used to fit the GPD model and estimate the yield at the specification point.

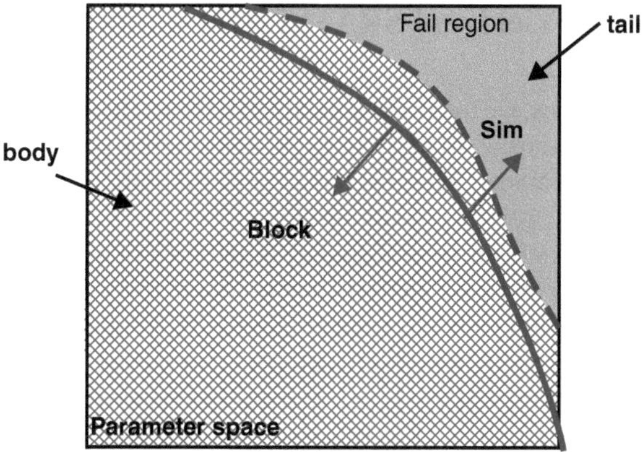

Fig. 5.5 Illustration of the statistical blockade method showing the tail and body regions in the statistical parameter space. The *dashed line* is the exact tail region boundary. The *solid line* is the relaxed boundary modeled by the classifier [27]

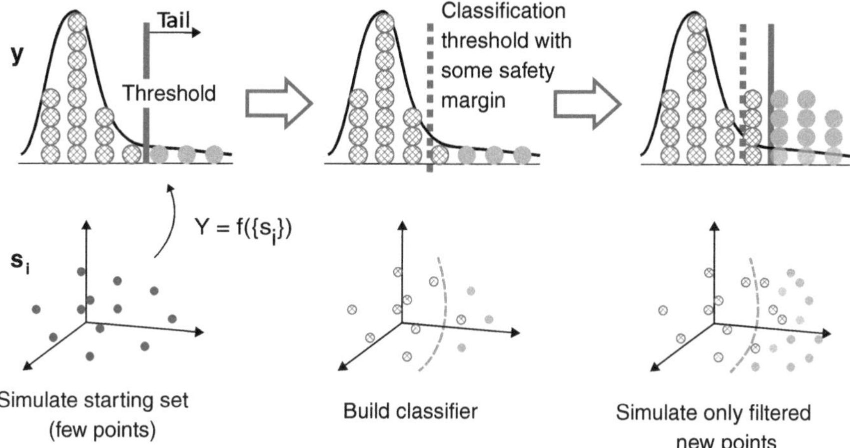

Fig. 5.6 Statistical Blockade steps: **a** simulate small set of points, **b** build classifier, **c** simulate only filtered points [27]

Statistical blockade has been applied to estimate SRAM write time and flop access time [27] as well as SRAM data retention voltage [29, 30]. In addition, improvements using recursive statistical blockade have been proposed [31] where statistical blockade starts with a classifier that uses a lower threshold, which is then used to estimate a higher threshold after every iteration. The initial use of a lower threshold classifier reduces the number of simulations for estimating the next higher threshold and these steps are repeated until the target threshold level is reached [31].

5.3 Read Access Yield and SRAM Performance Tradeoff

In the following sections, we focus on estimating yield loss due to read access failures, as this type of failure has a strong impact on determining performance and power consumption of the memory. Moreover, read access failure is the dominant failure mechanism at normal operating conditions [8].

Recently, there have been several studies in the area of statistical estimation of read access yield. A worst-case analysis that accounts for weak cells, and presents guidelines to achieve high yield has been proposed [4]. Another work models access failures by statistically accounting for bitcell read current variation as well as for the impact of access transistor leakage [7, 8]. These previous works determined memory yield for a given sense amplifier (SA) offset (i.e., a fixed value of bitlines differential voltage), assuming the worst-case for the SA offset, although statistical analysis is used for bitcell read current variations.

In this work, we generalize the access failures to statistically include the SA offset distribution. This helps SRAM circuit designers as it more accurately predicts access failures, and reduces the pessimism of assuming worst-case SA offset and worst-case bitcell. In addition, we include the impact of sensing window variation on yield. The proposed statistical yield estimation methodology accounts for bitcell read current variations, sense amplifier offset, and sensing window variations, as well as leakage from other bitcells on the same column. In particular, the proposed methodology helps answer the following questions for SRAM designers:

1. What is the maximum achieved performance (minimum sensing window) for a given yield requirement;
2. How much does a given SA offset improvement improve yield or performance (i.e., increasing SA area or changing SA topology);
3. How can the expected yield for memories having similar densities but different architectures be compared (i.e., yield for different memory options).

The read path in SRAM memories is typically a part of the critical path, which determines the memory access time (performance) [32]. Figure 5.7 shows the read path in an SRAM memory, consisting of an array of bitcells accessed using a shared SA. Each column of bitlines is selected using a column select multiplexer depending on the input addresses. Prior to selecting row and columns, the bitlines (BL and BLB) and sense lines (SL and SLB) are precharged to V_{DD}. Read operation begins by selecting the column using the PMOS pass-gate and activating the wordline (WL) of the selected row, as shown in Fig. 5.8. Depending on the stored data in the bitcell, one side of the bitlines begins to discharge the bitline capacitance using the bitcell read current (I_{read}). Therefore, a small differential voltage is generated at the inputs of the voltage SA (V_{SAin}). To ensure correct read operation, the SA is enabled using a control signal (SAEN) after a sufficient differential signal V_{SAin} is developed, which is amplified by the SA to a digital output level.

The delay difference between the activation of the WL and enabling of the SA is called the "read sensing window" ($t_{wl2saen}$), as shown in Fig. 5.8. $t_{wl2saen}$ directly

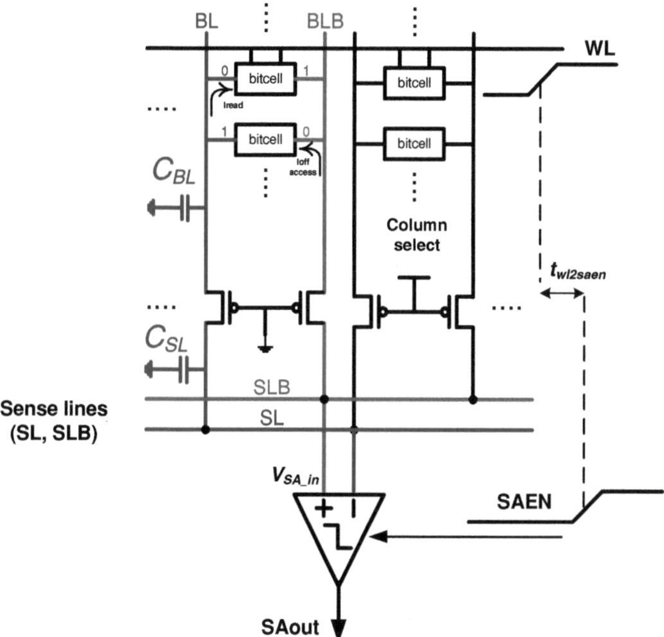

Fig. 5.7 Simplified SRAM read path

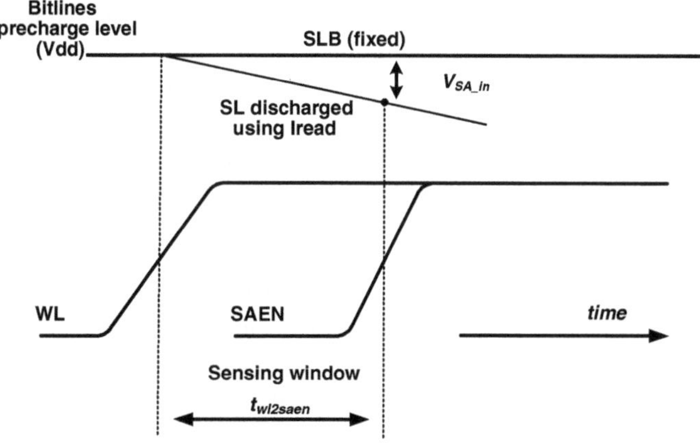

Fig. 5.8 Timing diagram for SRAM read operation

affects memory performance as it contributes a large percentage of the memory access time (~30 %) [33]. In addition, t_{wl2saen} affects dynamic power consumption. As t_{wl2saen} increases, the bitlines differential increases, which should be recovered by the precharge circuitry after each memory access cycle.[2] In the meantime, increasing

[2] Details about SRAM dynamic power are presented in Chap. 4.

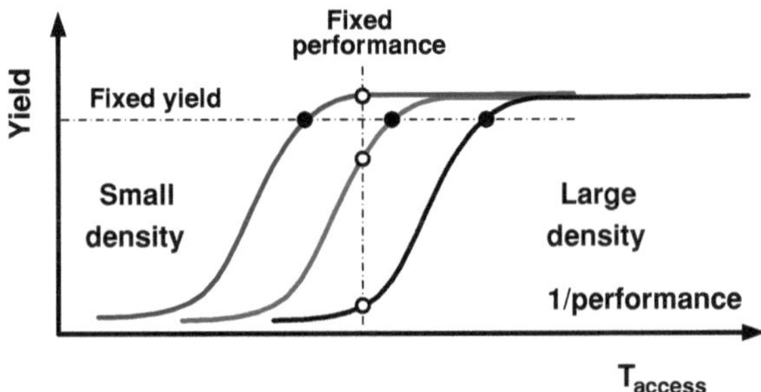

Fig. 5.9 Yield versus performance tradeoff for SRAM design

$t_{wl2saen}$ increases V_{SAin}, which reduces the probability of read failure due to SA input offset. Hence, it is always desirable to reduce $t_{wl2saen}$ as long as correct read operation is ensured. Therefore, determining $t_{wl2saen}$ involves a strong tradeoff between yield and performance/power for SRAM, which is one of the most important design decisions for memory designers.

Figure 5.9 shows typical yield[3] versus memory access time which is inversely proportional to the memory's maximum frequency (performance) for a given memory design and architecture. As the access time is shortened, by reducing $t_{wl2saen}$, memory yield drops. This is because decreasing $t_{wl2saen}$ decreases the SA input differential V_{SAin}, causing the SA to incorrectly sense the stored data. Moreover, memory density strongly affects the yield versus performance tradeoff. As the memory density increases (e.g., by using multiple instances of the same small memory macro), the whole curve shifts towards higher access time. Therefore, to achieve the same yield target similar to the smaller density, a larger memory will require a larger access time. If the performance is fixed, memory yield reduces for higher densities. This type of yield versus performance tradeoff is critical for memory design, and requires statistical simulation to accurately predict yield.

The statistical nature of SRAM failures requires statistical simulation techniques to account for these failure mechanisms early in the design cycle. Unfortunately, the problem of statistical design for memories is aggravated by circuit simulation speed and capacity limitations. Due to the large size of SRAM memories, it is very difficult to run Monte Carlo simulations for the whole memory. Even if the computational resources allow Monte Carlo simulation for the whole memory, an extremely large number of Monte Carlo runs is required, as discussed earlier in the chapter. In practice, SRAM designers typically use worst-case approaches to ensure high yield by designing for the worst-case bitcell for a given memory density [4]. However, this worst-case design technique lowers the performance as well as increases power consumption.

[3] This section focuses on read access yield. Therefore, we use the word yield to refer to read access yield.

Previous works in the area of statistical design for memories define a successful read access as the probability of having the bitlines reach a fixed voltage Δ_{min} for a fixed access (sensing) window $t_{wl2saen}$ [7, 8]. In [7, 8], although statistical analysis is performed on I_{read}, however, by assuming fixed Δ_{min} and $t_{wl2saen}$, this means that worst-case is assumed for the SA and the sensing window. In addition, previous models assumed that BL discharge could be coupled directly to the SA inputs. However, in reality, due to the on resistance of the PMOS column select device, the sense line is usually slower than the bitline discharge, and a longer time is required to achieve a certain differential voltage [32]. Therefore, the above-mentioned techniques are more appropriate for bitcell technology optimization, while a new access failure estimation methodology is required for memory circuit design that can account for different sources of access failures in a single statistical yield estimation flow. Statistical analysis is needed not only for bitcell V_{min} but for the product V_{min}. This means that the Monte Carlo simulation should go beyond the cell and include circuits and architecture components such as bitline length, word line pulse width, sense amplifier, write drivers, and other circuits that affect SRAM operation [34].

In the following sections, we explain how various factors affect access failures. Following that, we present a new read failure definition that is used in the proposed flow.

5.4 Modeling of Read Access Failures

As discussed in the previous section, the conventional worst-case analysis for read access limits the achieved performance and increases pessimism [4]. Nonetheless, a full statistical approach using circuit simulation is not an option due to practical limitations (simulation speed and capacity). In this work, we propose a full statistical analysis that overcomes the practical limitations of circuit simulation, instead, by using simple (but accurate enough) behavioral modeling for read access failure. The behavioral model's simplicity significantly improves simulation efficiency, since it can be used to run extensive Monte Carlo simulations. However, to derive a simple model for SRAM read failure, it is important to identify and account for the dominant sources that affect read operation.

There are four major contributors to read access failures in SRAM, and they are all strongly affected by process variations, as shown in Fig. 5.7:

1. bitcell read current variation;
2. SA input offset;
3. sensing window delay variation;
4. pass-gate (access transistor) leakage.

5.4.1 Read Current and Sensing Slope Variations

Due to the small size of SRAM bitcell and the inverse relation between transistor variation and device area [35, 36], bitcell read current I_{read} shows large within-die (WID) variations [3, 8], and typically follows a normal distribution. From a memory design point of view, I_{read} determines the time required to develop enough differential signal before enabling the SA. I_{read} variation is one of the largest sources of parametric yield loss in memories [8].

As mentioned earlier, SL are discharged using the bitcell I_{read}. However, the SL discharge rate is slower than that for bitlines due to the ON resistance of the column select device (PMOS), as shown in Fig. 5.7, which adds RC delay at the SA input [32]. The sense lines discharge slope can be defined as $K_{\text{eff}} = |\Delta V_{\text{SL}}/\Delta t|$. K_{eff} is proportional to I_{read} [32], therefore, the statistical variation in I_{read} causes similar variation in K_{eff}, and $\frac{\sigma}{\mu}|_{K_{\text{eff}}} = \frac{\sigma}{\mu}|_{I_{\text{read}}}$.

Due to random variations in bitcells transistors, each bitcell has two values of I_{read} depending on the stored data (whether it is 0 or 1). I_{read} path is different for the case of read 0 or read 1 as shown in Fig. 5.10. This difference in I_{read} values occurs since each transistor in the bitcell experiences different value of random WID variation. Therefore, I_{read} for read 0 and read 1 cases are statistically independent.

5.4.2 Sense Amplifier Variations

An SA is typically used to amplify the small differential voltage on the bitlines (∼100 mv) to a digital output level [4]. Figure 5.11 shows one of the most widely used SAs due to its fast decision time. This SA is also called a decoupled SA (DSA) since the inputs and outputs of the SA are separated [37, 38]. The decision threshold of the SA is ideally zero. That is, if $V_{\text{SAin}} > 0$, the output is high, and the output is low, if $V_{\text{SAin}} < 0$. The amount by which the threshold point shifts is called the input offset [39].

SAs are extremely sensitive to WID variations (mismatch) [38, 40, 41], which cause SAs to have offset voltages that affect the accuracy of read operation. In addition, systematic variations due to asymmetric layout can increase the SA input offset. One way to reduce SA input offset is to increase the size of input devices [35, 38]. Due to the strict limitations on area in memory design, the SA area-mismatch trade-off is difficult because the SA should pitch-match the accessed bitcells. Therefore, the specification on SA offset is an important metric for memory designers.

Monte Carlo transient simulation is usually used to estimate the input offset distribution of an SA [4, 42, 43]. Typically, the SA input offset can be modeled using a Gaussian distribution with a mean of zero and standard deviation of $\sigma_{V_{\text{SA offset}}}$ as shown in Fig. 5.12. The SA input offset can be modeled as a noise voltage source connected at the input of an ideal SA, as shown in Fig. 5.13, where the voltage source follows the normal distribution of the $V_{\text{SA offset}}$. Methods to reduce the SA

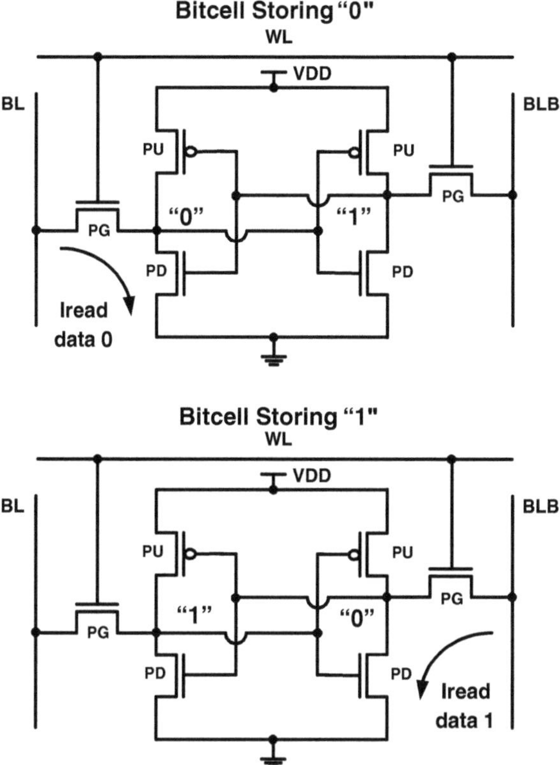

Fig. 5.10 6T bitcell showing the different read current paths for read 0 and read 1

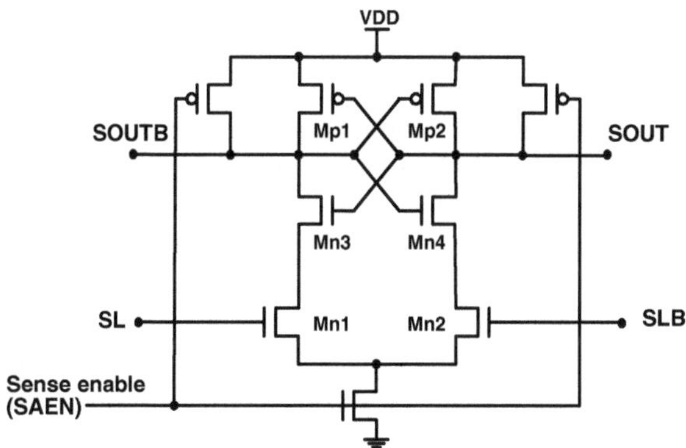

Fig. 5.11 Current latch sense amplifier (CLSA) [37, 38]

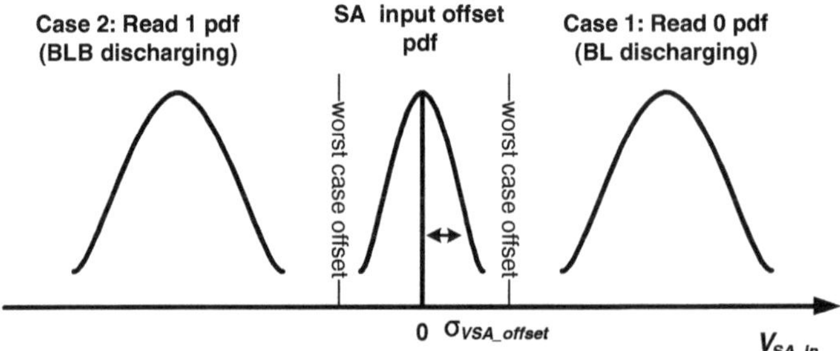

Fig. 5.12 SA input offset and read 0/1 distributions

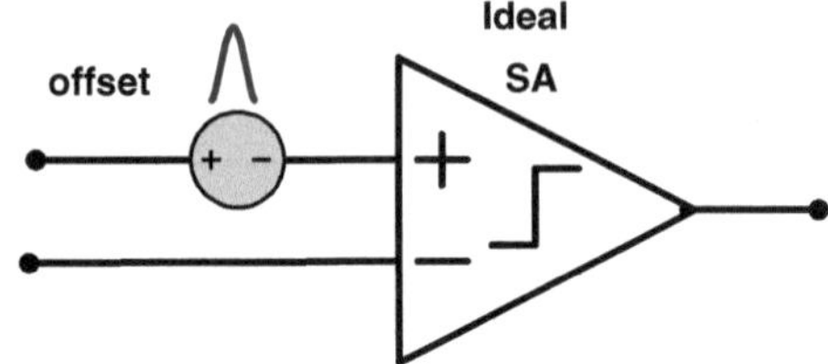

Fig. 5.13 Modeling the SA input offset as a noise source connected at the input terminal of SA

offset range from sizing [38, 44, 45], optimizing sense enable signal slew [46] and different SA topologies [47].

As mentioned in Sect. 5.4.1, I_{read} for read 0 and read 1 cases are statistically independent. However, a shared SA is used to sense both states as shown in Fig. 5.7. Therefore, two distributions are used to model the SA input voltage (proportional to I_{read}), while one distribution is used to model the SA input offset (V_{SAoffset}) as shown in Fig. 5.12. In a worst-case design scenario, the minimum sensing voltage V_{SAin} is required to be larger than the worst-case SA offset (as shown in Fig. 5.12). This is a pessimistic approach because the probability of having the slowest bitcell accessed using the SA suffering the largest offset is very small.

Another source that can increase read access failures is dynamic noise coupling at the SA inputs. Due to the small differential signal developed on the SL, an aggressor located near the SA may couple large noise at the SA input which can affect the accuracy of read operation. This situation is exacerbated if a weak bitcell is selected, the read sensing window is reduced, or if the noise occurs just before the SA is enabled. However, modeling of this dynamic noise component is very complex, as it strongly depends on the layout of the SA and SL, as well as the timing of the aggressors relative to the SA enable signal (SAEN). Moreover, capacitance imbalance in the SA can cause large increase in the SA offset [48]. Nevertheless, by using layout noise shielding techniques and highly symmetric SA layout styles, the impact of this component can be minimized.

There are other factors that affect SA operation, such as SA delay and intrinsic noise. The SA delay depends on the SA input differential, where as the input differential decreases, the SA delay increases exponentially as shown in Fig. 5.14. This

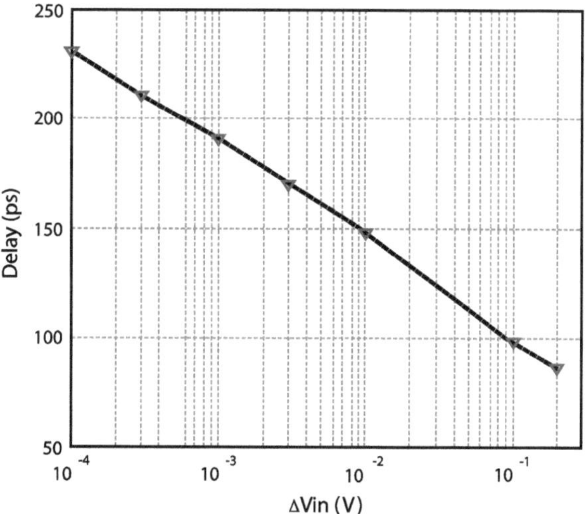

Fig. 5.14 Sense amplifier delay versus the SA input differential [47]

increase may cause a read access failure since a larger SA delay increases the access time of the memory.

Another factor that affects the SA operation is the intrinsic device noise [4] (thermal and flicker noise) [47, 50]. Device noise causes the SA to fail at different times randomly, which increases the bit error rate (BER) as shown in Fig. 5.15. SA offset variation due to noise can be extracted from BER measurement, which in Fig. 5.15 shows that $\sigma_{V_{\mathrm{SA\,offset}}}$ due to device noise is 0.8 mV. However, since the device noise component is much lower than the static offset due to RDF and LER, it is typically ignored in SRAM design. Nevertheless, the intrinsic device noise component can become the dominant source of SA failure for SA topologies that employ offset cancelation approaches.

5.4.3 Sensing Window Variations

As mentioned earlier, the read sensing window t_{wl2saen} is an important parameter for correct read operation. In memory design, a centralized control block (timer circuit) is used to generate the timing for all the critical signals for memory operation—which include WL and SAEN signals [4]. To ensure good tracking with PVT variations, similar transistor sizes are usually used in the two logic paths [4, 51]. However, due to random WID variations, the delay in these paths will vary stochastically [5, 21, 52]. Therefore, the sensing window will vary around its mean value (as shown in Fig. 5.16).

[4] For detailed analysis on transistor intrinsic noise, the reader is referred to [49].

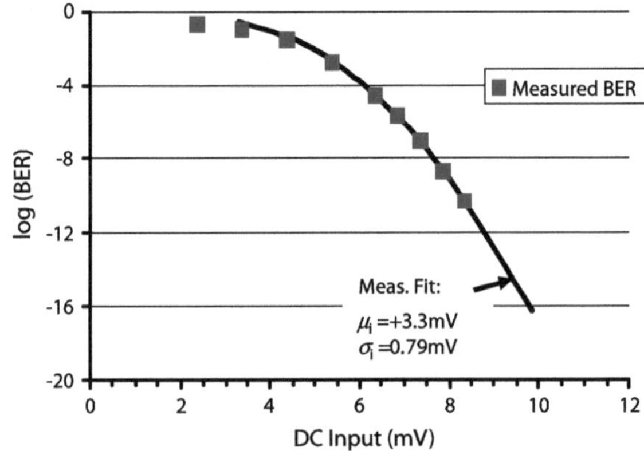

Fig. 5.15 Measured SA bit error rate (BER) due to intrinsic device noise [50]

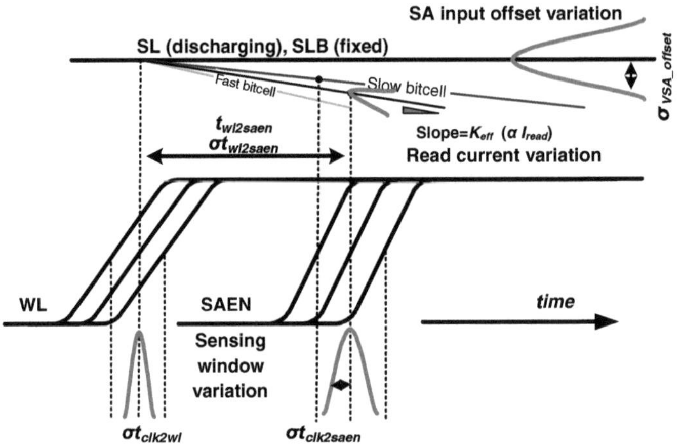

Fig. 5.16 Sources of variation affecting access failures

To estimate the impact of delay variation, we assume that the number of logic stages between internal CLK to WL and SAEN is m and n stages, respectively, as shown in Fig. 5.17. For the sake of simplicity, we further assume that the delay of each path can be modeled as a chain of inverters, with t_d being the delay of one stage. In an ideal scenario with no random WID variations, t_{wl2saen} can be computed as $(n-m)t_d$. However, due to random variations, t_{wl2saen} will vary, and its distribution typically assumed to be Gaussian [5, 6, 21]. Therefore, the mean and variance of t_{wl2saen} can be computed as $\mu_{t_{\text{wl2saen}}} = (n-m)\mu_{t_d}$ and ${\sigma^2}_{t_{\text{wl2saen}}} = (n^2+m^2){\sigma^2}_{t_d}$, respectively, where ${\sigma^2}_{t_d}$ is the variance of one delay stage. In the case of memories, n and m are comparable, where $n-m$ determines the nominal t_{wl2saen}. However, the sensing window variations is large since the variation in each logic path adds up to the t_{wl2saen} variation (n^2+m^2 term). Note also that the variation of $\sigma_{t_{\text{wl2saen}}}$ increases as n and m increase even if $n-m$ is constant (i.e., for a fixed t_{wl2saen}

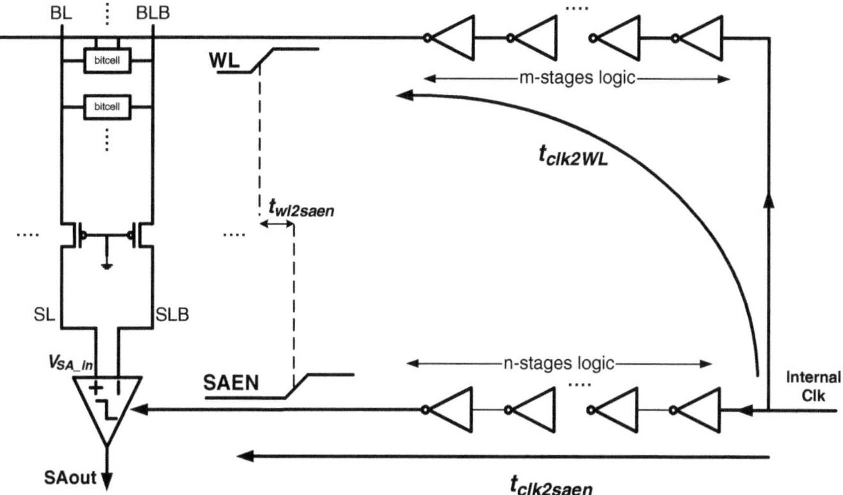

Fig. 5.17 Timing delay variation between WL and SAEN paths

delay). This implies that as the memory size increases and more logic stages are required in the CLK to WL and SAEN paths, this effect becomes more severe. This variation in sensing window can reduce the SA input voltage, which increases access failure probability—especially at low supply voltages, since $\frac{\sigma}{\mu}|_{t_{\text{wl2saen}}}$ increases due to reduced headroom [5].

While a chain of inverters can be used to qualitatively explain the importance of accounting for read window variations, a more comprehensive delay variation analysis is required to account for different logic gates, input slews, and fanouts in the CLK to WL and CLK to SAEN paths. In this work, we use Monte Carlo simulation to determine $\mu_{t_{\text{wl2saen}}}$ and $\sigma_{t_{\text{wl2saen}}}$. Nevertheless, statistical timing analysis [6] can also be used for the same purpose.

5.4.4 Pass-Gate Leakage

Bitcell pass-gate (access) device leakage also reduces the SA input differential due to subthreshold leakage from the other side of the bitlines [8]. The worst-case sensing occurs when all the unselected bitcells on the column store the opposite data from that on the selected bitcell, as shown in Fig. 5.18. Pass-gate transistor leakage determines the upper limit of the number of bitcells per column. This effect is usually important in high performance memories due to the high leakage (low V_{th}) of the pass-gate device. The effective read current can be calculated as [8]:

$$\mathbf{I_{read,eff}} = \mathbf{I_{read}} - \sum_{i=1}^{N_c-1} \mathbf{I_{off,PG,i}} \tag{5.13}$$

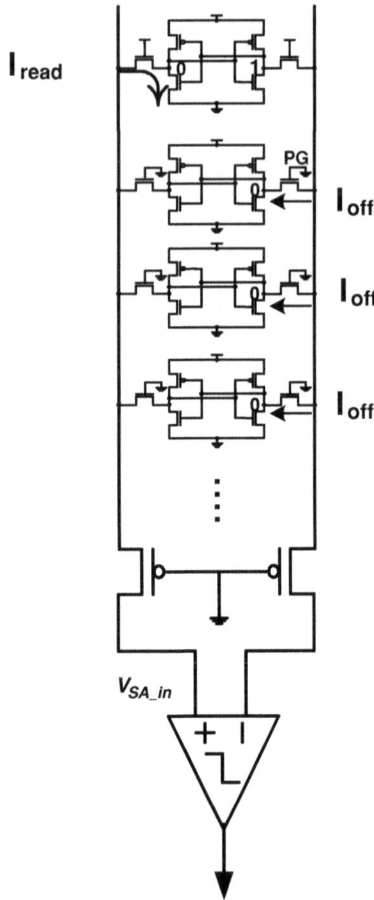

Fig. 5.18 Impact of pass-gate leakage on the SA's input differential voltage (V_{SAin}). Pass-gate leakage from adjacent bitcells on the same column reduces the effective read current for the selected bitcell

where $\mathbf{I_{read,eff}}$ is the effective read current for a bitcell after accounting for the pass-gate leakage for adjacent bitcells on the same column, N_c is the number of bitcells per column, and $\mathbf{I_{off,PG,i}}$ is the pass-gate leakage for one bitcell.[5]

Due to the exponential dependence of subthreshold leakage on V_{th} variations, it is important to statistically calculate the total leakage of all pass-gate devices. Assuming subthreshold leakage is the dominant leakage source, and that there are a large number of bitcells on the same column, it can be shown that [53, 54]:

$$\begin{aligned}\sum_{i=1}^{N_c-1} \mathbf{I_{off,PG,i}} &= (N_c - 1)\,\mu_{I_{\mathrm{off,PG}}} \\ &\approx (N_c - 1)\,I_{\mathrm{off,PG}} \left(1 + \frac{\ln^2(10)}{2}\left(\frac{\sigma_{V_{\mathrm{th}}}}{S}\right)^2\right)\end{aligned} \tag{5.14}$$

[5] A **bold** symbol is used to indicate a random variable.

where $\mathbf{I_{off,PG}}$ is the nominal PG leakage (assuming there is no V_{th} variation), $\sigma_{V_{th}}$ is the variation in V_{th} for the pass-gate device due to random WID variations, and S is the subthreshold slope. From Eq. (5.14), it is clear that larger $\sigma_{V_{th}}$ increases the total PG leakage, which reduces the effective read current as shown in Eq. (5.13).

5.5 Proposed Yield Estimation Flow

Figure 5.16 shows a simplified timing diagram for an accessed bitcell including critical signals such as WL, SAEN, and SL. Also shown in the figure are the statistical variations of different components that affect the probability of access failure, described in the (Sect. 5.4). When the WL is enabled, SL begins discharging, and the slope of SL discharge varies statistically depending on I_{read} variations as well as leakage from other bitcells (Sect. 5.4.4).

For the SA, the offset voltage distribution is superimposed on the V_{DD} (precharge level). As explained in Sect. 5.4.2, SA offset distribution is centered around zero (typically small asymmetry), as shown in Fig. 5.12, which means that some SAs will have positive offsets while others will have negative offsets. Note that positive offset will increase the failure probability of reading a 0 and reduce the failure probability of reading a 1, and vice versa. Therefore, in order to account for SA offset statistically in yield estimation, both read 0 and read 1 cases need to be addressed.

In addition to I_{read} and SA offset, $t_{wl2saen}$ variation can affect access failure probability, as described in Sect. 5.4.3. As shown in Fig. 5.16, if $t_{wl2saen}$ decreases due to statistical variations, V_{SAin} decreases, and hence the probability of access failure increases.

In order to estimate SRAM yield, it is important to statistically account for all the above-mentioned contributors to failure probability in the same flow. Therefore, we define access failure for a certain bitcell as follows: for the read 0 case, the probability of access failure is the probability that the SA input voltage V_{SAin} is less than the SA input offset $V_{SAoffset}$ of that particular SA. Note that we do not assume a fixed value of SA offset as in [7, 8]. Instead, the SA offset follows the normal distribution from Monte Carlo simulations. Moreover, V_{SAin} needs to be computed statistically since it is a function of the distributions of bitcell I_{read} and $t_{wl2saen}$.

Therefore, the probability of access failure for bitcell $P_{AF,cell}$ in case of reading a 0 can be expressed mathematically as follows:

$$\begin{aligned} P_{AF,cell,\,read0} &= P(\mathbf{V_{SAin}} - \mathbf{V_{SAoffset}} < 0) \\ &= P(\mathbf{K_{eff0}} \mathbf{t_{wl2saen}} - \mathbf{V_{SAoffset}} < 0) \end{aligned} \tag{5.15}$$

where the sense slope $\mathbf{K_{eff0}}$, $\mathbf{t_{wl2saen}}$ and $\mathbf{V_{SAoffset}}$ are all random variables following a normal distribution, as explained in Sect. 5.4. A similar expression can be derived for the probability of read 1 access failure. To account for pass-gate leakage, K_{eff} distribution can be calculated as:

$$\begin{aligned}
&\mathbf{K_{eff0}} \text{ follows } \mathcal{N} \sim (\mu_{K_{\text{eff}}}, \sigma^2_{K_{\text{eff}}}) \\
&\mu_{K_{\text{eff}}} = \left|\frac{\Delta V_{\text{SL}}}{\Delta t}\right| \left(1 - \frac{(N_c - 1)\mu_{I_{\text{off,PG}}}}{\mu_{I_{\text{read}}}}\right) \\
&\sigma_{K_{\text{eff}}} = \left|\frac{\Delta V_{\text{SL}}}{\Delta t}\right| \frac{\sigma_{I_{\text{read}}}}{\mu_{I_{\text{read}}}}
\end{aligned} \tag{5.16}$$

Table 5.2 summarizes the inputs for the read failure model. For the typical memory architecture shown in Fig. 5.19, using the proposed access failure definition in Eq. (5.15), the flow for read access yield computation is implemented as follows (as shown in the flowchart in Fig. 5.20):

1. From the memory architecture (density, word length, number of columns, muxing), find the number of banks (N_{banks}), SAs per bank ($N_{\text{SA-bank}}$) and number of bitcells accessed by one SA ($N_{\text{bits-SA}}$);
2. Initialize the chip counter (number of Monte Carlo runs);
3. Generate one sample of t_{wl2saen}: $\mathcal{N} \backsim (\mu_{t_{\text{wl2saen}}}, \sigma^2{}_{t_{\text{wl2saen}}})$[6];
4. Generate one sample SA input offset distribution: $\mathcal{N} \backsim (\mu_{V_{\text{SAoffset}}}, \sigma^2{}_{V_{\text{SAoffset}}})$;
5. Generate $2N_{\text{bit-bank}}$ samples of K_{eff} normal distribution using Eq. (5.16) to represent the read 0 and read 1 sensing slope distributions (K_{eff0}, K_{eff1});
6. Failure calculation step: loop on all the bitcells accessed using this particular SA. Check the following fail conditions for each bitcell;

   - Read 0 fail: $K_{\text{eff0}}\, t_{\text{wl2saen}} - V_{\text{SAoffset}} < 0$
   - Read 1 fail: $K_{\text{eff1}}\, t_{\text{wl2saen}} - V_{\text{SAoffset}} > 0$
   - Count the number of read failures.

7. Repeat all the above steps for all SAs per bank ($N_{\text{SA-bank}}$);
8. Repeat all the above steps for all banks (N_{banks});
9. Repeat all the above steps for a large number of chips (Monte Carlo runs), and count the number of failing chips;
10. Calculate the yield based on the number of chips that can correctly be accessed for the read 0 and 1 cases, where yield = (number of passing chips)/(total number of chips).

While the above steps focused on WID variation, the proposed methodology can be easily extended to account for die-to-die (D2D) variations. This can be done by including the statistical distributions of D2D variations and including a large number of D2D samples for each run at the chip level (i.e., in step 9 shown above). However, this will also require pre-characterization of the failure model parameters (Table 5.2) at all the D2D sample points.

[6] Here, we assume that a bank contains one control block that generates WL and SAEN signals as shown in Fig. 5.19. Nevertheless, different types of banking styles can be easily included in the flow.

Table 5.2 Read failure model inputs for the proposed statistical yield estimation methodology

	Parameter	Estimation technique
I_{read} variation	$\mu_{I_{\mathrm{read}}}, \sigma_{I_{\mathrm{read}}}$	DC MC[a] simulation for the bitcell
Sensing slope	$\lvert\frac{\Delta V_{\mathrm{SL}}}{\Delta t}\rvert$	Transient simulation at nominal conditions (no variations) for bitlines discharge
SA input offset	$\mu_{V_{\mathrm{SA\,offset}}}, \sigma_{V_{\mathrm{SA\,offset}}}$	Transient MC simulation for SA
Sensing window variation	$\mu_{t_{\mathrm{wl2saen}}}, \sigma_{t_{\mathrm{wl2saen}}}$	Transient MC simulation for WL and SAEN paths
PG leakage	$\mu_{I_{\mathrm{off,PG}}}$	DC MC simulation for PG transistors

[a]*MC* Monte Carlo

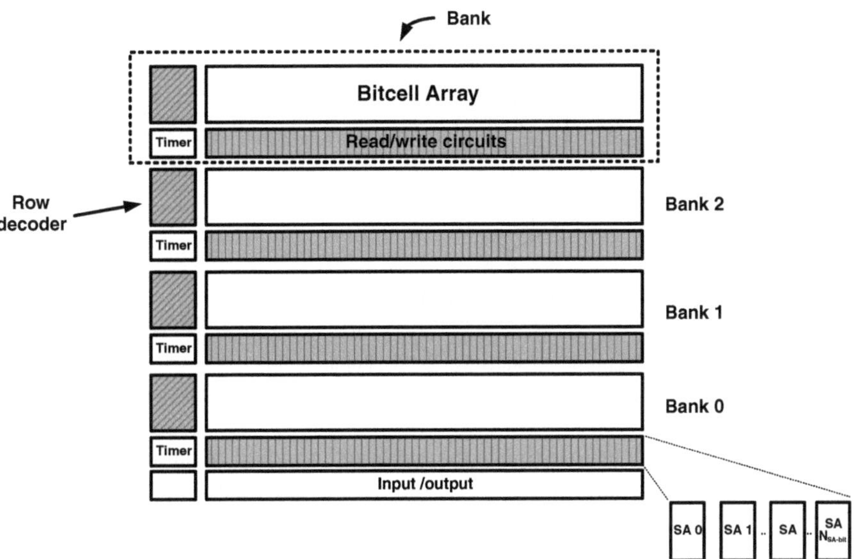

Fig. 5.19 Typical SRAM architecture used in the proposed statistical yield estimation flow

5.6 Experimental Results

The proposed yield estimation methodology was verified using a 1 Mb SRAM design fabricated in an industrial 45 nm CMOS technology. Prior to running the proposed yield estimation flow, a characterization step is required to compute the inputs for the proposed flow (Table 5.2). However, this characterization step is not computationally expensive due to the reduced number of circuit elements for these simulation setups. In addition, these extracting the inputs for the failure model is an part of any SRAM design and should be readily available even when using worst-case analysis.

Characterization for the different components of yield failures was performed as shown in Fig. 5.21 for different conditions. I_{read} was characterized using DC Monte

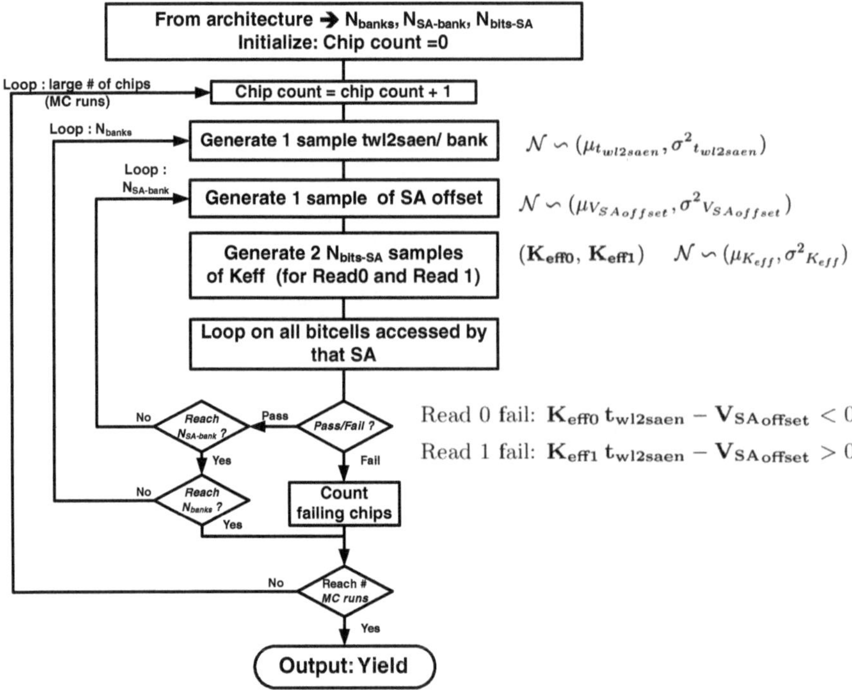

Fig. 5.20 Flowchart of the proposed statistical yield estimation flow

Carlo SPICE simulations to estimate the mean and standard deviation. Figure 5.22 shows the I_{read} histogram for 100 k MC runs. As expected, I_{read} follows a normal distribution which is used to extract $\mu_{I_{\text{read}}}$ and $\sigma_{I_{\text{read}}}$. We also ran the simulations at various voltage and temperature conditions to determine how they affect $\sigma/\mu|_{I_{\text{read}}}$, as shown in Fig. 5.23. Note that $\sigma/\mu|_{I_{\text{read}}}$ reaches 20 % at low V_{DD} and low temperature, which shows the strong impact of process variations on bitcell I_{read}.

The sensing slope ($|\frac{\Delta V_{SL}}{\Delta t}|$) was extracted by running a transient SPICE simulation for bitline discharge rate. Monte Carlo simulation is not required in this case since I_{read} variation estimated from the first step is used to estimate K_{eff} variation, as shown in Eq. (5.16). SA offset distribution was simulated using Monte Carlo transient simulation as shown in Fig. 5.24, which compares the simulated/modeled cumulative distribution functions (CDF) for the SA input offset, indicating that the Gaussian model for SA offset matches the simulated data. We also simulated the impact of V_{DD} and temperature on SA input offset, as shown in Fig. 5.25; both strongly the $\sigma_{V_{\text{SA offset}}}$.

The distribution of sensing window t_{wl2saen} was estimated using Monte Carlo transient simulation on the WL and SAEN paths of the memory. Figure 5.26 shows the impact of V_{DD} and temperature on t_{wl2saen} variation. Similar to I_{read} variation, t_{wl2saen} variation increases as V_{DD} or temperature is reduced.

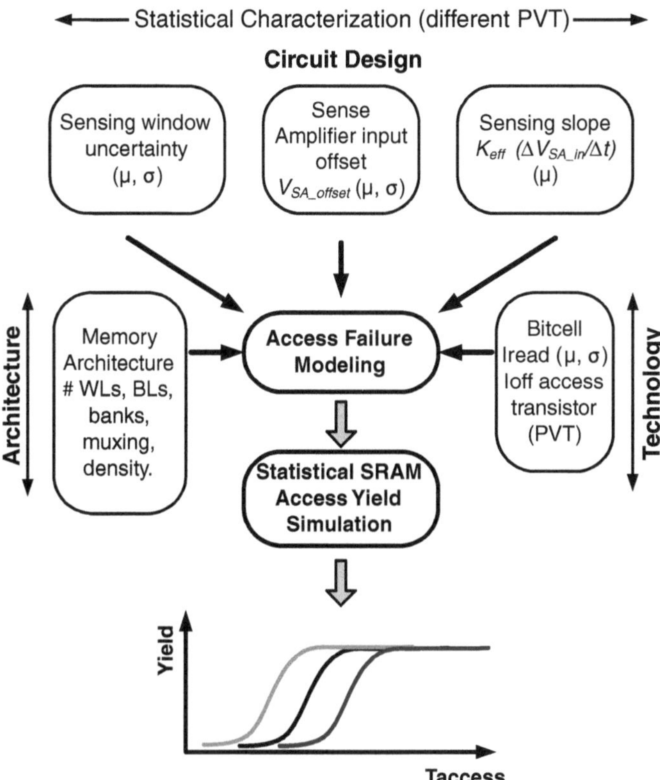

Fig. 5.21 Yield estimation flow

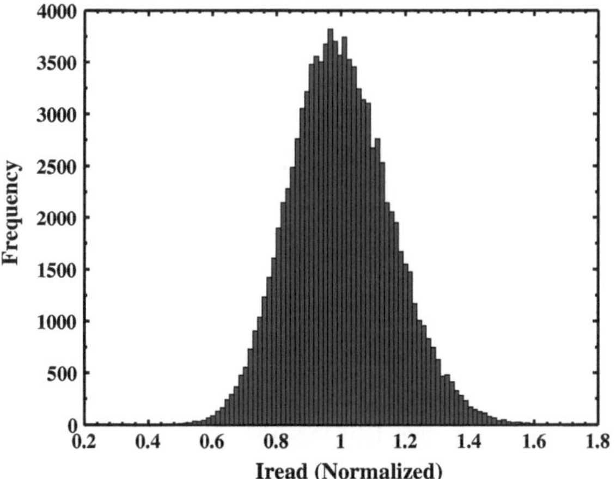

Fig. 5.22 I_{read} histogram from Monte Carlo simulation (100 k runs)

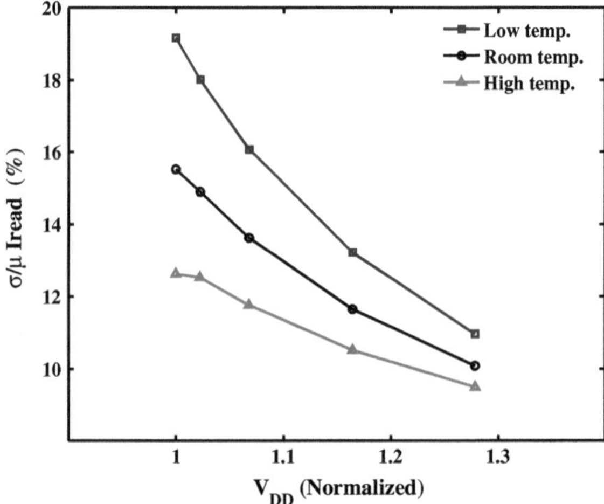

Fig. 5.23 Characterization results for bitcell I_{read} variation using DC Monte Carlo simulation

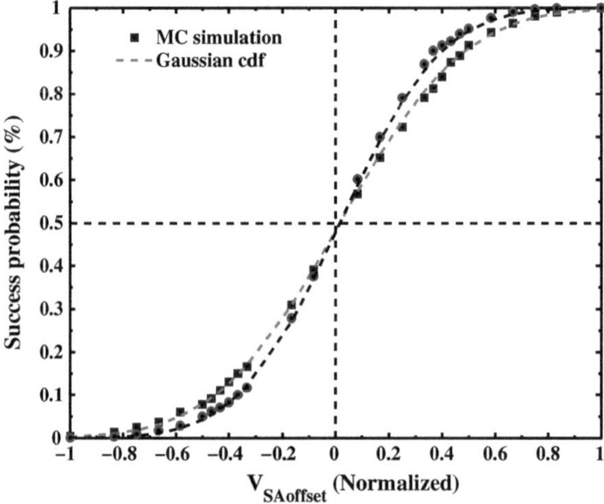

Fig. 5.24 Simulated SA offset distribution using transient Monte Carlo Analysis for different conditions. SA input offset follows a normal distribution as shown by the cdf (cumulative distribution function) of simulation and model

For pass-gate leakage, DC Monte Carlo simulation was used to estimate $\mu_{I_{\text{off,PG}}}$. Figure 5.27 shows the pass-gate leakage histogram. Notice how the distribution follows a lognormal shape. From simulation results, it was found that $\mu_{I_{\text{off,PG}}}$ is 1.3X larger than nominal pass-gate leakage. Figure 5.28 shows how $\mu_{I_{\text{off,PG}}}$ varies at different temperatures and V_{DD} conditions. Since pass-gate leakage is dominated by subthreshold leakage, it is clear that the impact of temperature is the dominant.

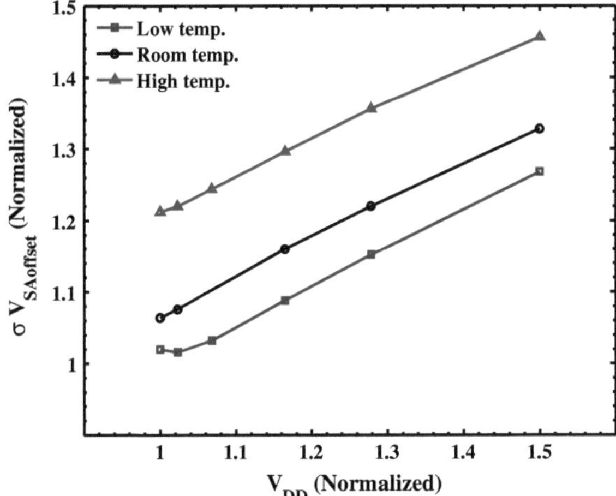

Fig. 5.25 Simulated results for $\sigma_{V_{SA_{offset}}}$ versus V_{DD} at different temperatures using Monte Carlo simulation

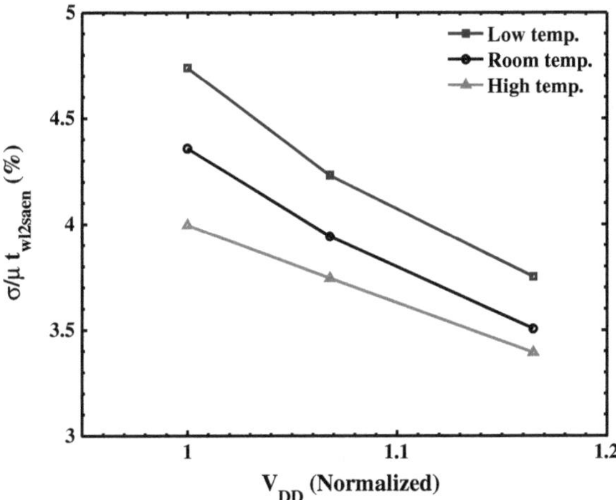

Fig. 5.26 Simulated results sensing window variation versus V_{DD} at different temperatures using Monte Carlo simulation

After generating the characterization data, and entering the memory architecture information, the statistical yield simulation described in Sect. 5.5 was executed using Matlab. Figure 5.29 shows the measured yield from the 1 Mb memory compared to the simulation for different supply voltage conditions. Good agreement between silicon and simulation results validate the accuracy of the proposed methodology. For these simulation results, 1,000 chips of the 1 Mb memory were simulated using the proposed flow. All bitcells were tested for read 0 and read 1 fails. Yield estimation

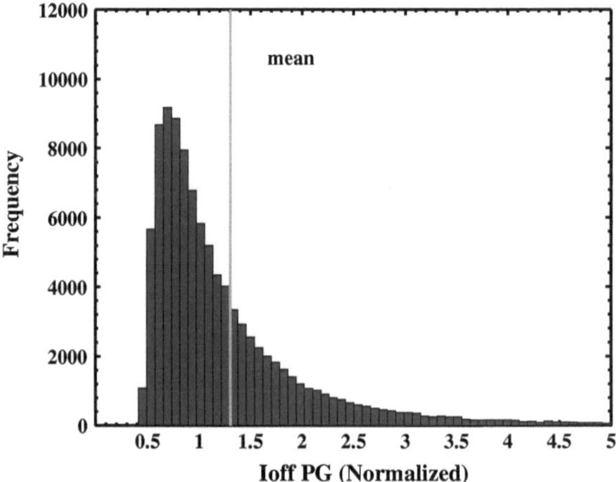

Fig. 5.27 Pass-gate leakage distribution

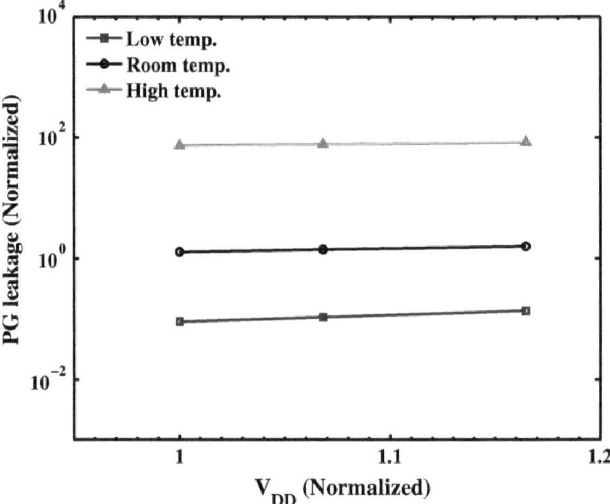

Fig. 5.28 Characterization results for $\mu_{I_{\text{off,PG}}}$ versus V_{DD} at different temperatures using Monte Carlo simulation

for the proposed methodology takes less than 30 min to generate the results shown in Fig. 5.29 using a 3 GHz PC with 1.5 GB of memory which shows the efficiency of the proposed flow. The simulation results in Fig. 5.29 can be used to explore the critical tradeoff between performance and yield requirement.

Comparing the proposed statistical yield estimation flow to the worst-case analysis allows calculation of the improvement in performance through the proposed flow. This is shown in Fig. 5.30, where in the worst-case approach, the worst bitcell is assumed to occur with the SA having the largest offset and the smallest sensing window. Also shown is the minimum t_{wls2saen} estimated by the statistical design

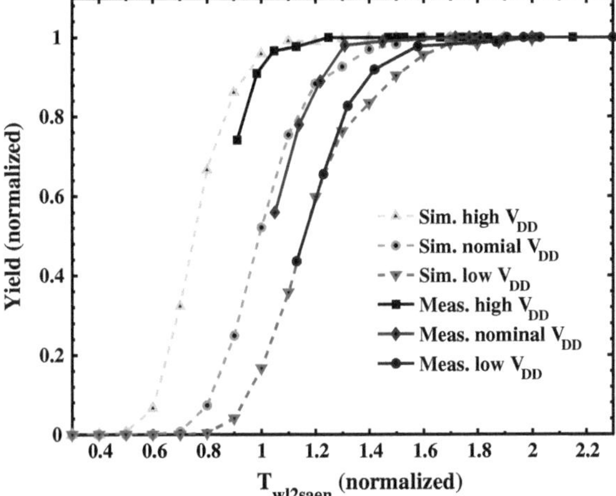

Fig. 5.29 Comparison between simulation results using the proposed yield estimation methodology and the measured access yield for a 1 Mb memory in 45 nm technology

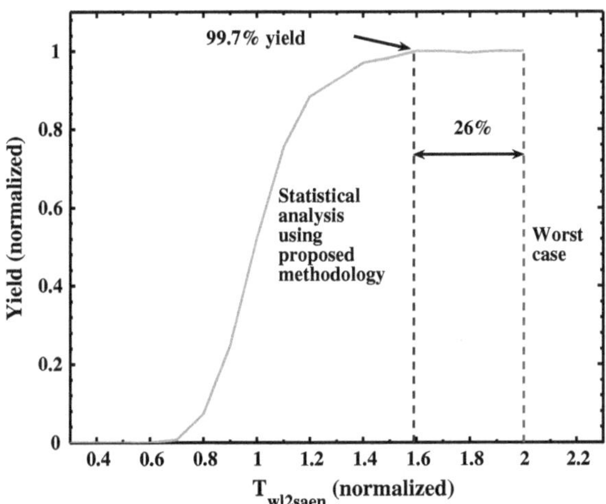

Fig. 5.30 Comparison between the proposed statistical yield estimation methodology and worst-case analysis

approach that meets a yield of 99.7 %. The statistical design enables reducing $t_{wls2saen}$ by 26 %, which translates into higher memory performance. This translates to 8% faster access time assuming $t_{wls2saen}$ is 30 % of access time [33]. Simultaneously, the memory read power consumption also reduces because of reduced differential voltage on the bitlines.

The performance benefit of using statistical methodology versus worst-case approaches increases as memory density increases. It is expected that the difference between the two approaches would increase with scaling due to the continuous increase in process variations as well as the higher SRAM density requirements. Statistically accounting for various components of read failures will become even more important in the future to avoid the pessimism in worst-case approaches.

5.7 Summary

The large increase in statistical variations in nanometer technologies is presenting huge challenges for SRAM designers. An important part of the SRAM design and optimization is the estimation of failure probability due to random variations. In this chapter, simulation methods for estimating SRAM failure probability were discussed including conventional Monte Carlo as well as newer methods such as IS, statistical blockade, and MPFP. As a case study for statistical yield estimation, a methodology for statistical estimation of access yield was presented. The flow accounts for the impact of bitcell read current variation, SA offset distribution, timing window variation, and leakage variation on read failure. The methodology overcomes the pessimism in worst-case design techniques that are usually used in SRAM design. The methodology is verified using measured yield data from a 1 Mb memory in an industrial 45 nm technology. The proposed statistical SRAM yield estimation methodology allows early yield prediction in the design cycle, which can be used to tradeoff yield, performance, and power requirements for SRAM.

References

1. The International Technology Roadmap for Semiconductors (ITRS), http://public.itrs.net
2. H. Masuda, S. Ohkawa, A. Kurokawa, M. Aoki, Challenge: variability characterization and modeling for 65- to 90-nm processes, in *Proceedings of IEEE Custom Integrated Circuits Conference*, 2005, pp. 593–599
3. A. Asenov, A. Brown, J. Davies, S. Kaya, G. Slavcheva, Simulation of intrinsic parameter fluctuations in decananometer and nanometer-scale MOSFETs. IEEE Trans. Electron Devices *50*(9), 1837–1852 (2003)
4. R. Heald, P. Wang, Variability in sub-100 nm SRAM designs, in *Proceedings of International Conference on, Computer Aided Design*, 2004, pp. 347–352
5. M. Eisele, J. Berthold, D. Schmitt-Landsiedel, R. Mahnkopf, The impact of intra-die device parameter variations on path delays and on the design for yield of low voltage digital circuits. IEEE Trans. Very Large Scale Integr. Syst. **5**(4), 360–368 (1997)
6. A. Srivastava, D. Sylvester, D. Blaauw, *Statistical Analysis and Optimization for VLSI: Timing and Power (Series on Integrated Circuits and Systems)* (Springer, New York, 2005)
7. K. Agarwal, S. Nassif, Statistical analysis of SRAM cell stability, in *Proceedings of the 43rd Annual Conference on Design Automation DAC '06*, 2006, pp. 57–62
8. S. Mukhopadhyay, H. Mahmoodi, K. Roy, Statistical design and optimization of SRAM cell for yield enhancement, in *Proceedings of International Conference on, Computer Aided Design*, 2004, pp. 10–13
9. M. Orshansky, S. Nassif, D. Boning, *Design for Manufacturability and Statistical Design: A Comprehensive Approach* (Springer-Verlag, Secaucus, 2006)

10. F. Gong, Y. Shi, H. Yu, L. He, Parametric yield estimation for SRAM cells: concepts, algorithms and challanges, in *Design Automation Conference, Knowledge Center Article*, 2010
11. X. Li, J. Le, L.T. Pileggi, *Statistical Performance Modeling and Optimization*, vol. 1 (Now Publishers Inc., Hanover 2006)
12. A. Bhavnagarwala, X. Tang, J. Meindl, The impact of intrinsic device fluctuations on CMOS SRAM cell stability. IEEE J. Solid-State Circuits **36**(4), 658–665 (2001)
13. S. Mukhopadhyay, H. Mahmoodi, K. Roy, Modeling of failure probability and statistical design of SRAM array for yield enhancement in nanoscaled CMOS. IEEE Trans. Comput. Aided Des. Integr. Circuits Syst. **24**(12), 1859–1880 (2005)
14. B. Calhoun, A. Chandrakasan, Analyzing static noise margin for sub-threshold SRAM in 65 nm CMOS, in *Proceedings of the 31st European, Solid-State Circuits Conference, ESSCIRC* 2005, pp. 363–366, Sept 2005
15. B. Calhoun, A. Chandrakasan, Static noise margin variation for sub-threshold SRAM in 65-nm CMOS. IEEE J. Solid-State Circuits **41**(7), 1673–1679 (2006)
16. M.H. Abu-Rahma, K. Chowdhury, J. Wang, Z. Chen, S.S. Yoon, M. Anis, A methodology for statistical estimation of read access yield in SRAMs, in *Proceedings of the 45th Conference on Design Automation DAC '08*, 2008, pp. 205–210
17. J. Wang, S. Yaldiz, X. Li, L. Pileggi, SRAM parametric failure analysis, in *46th ACM/IEEE, Design Automation Conference, DAC '09*, pp. 496–501, July 2009
18. A. Papoulis, *Probability, Random Variables, and Stochastic Processes*, 3rd edn. (McGraw-Hill, New York, 1991)
19. E. Grossar, M. Stucchi, K. Maex, W. Dehaene, Read stability and write-ability analysis of SRAM cells for nanometer technologies. IEEE J. Solid-State Circuits **41**(11), 2577–2588 (Nov. 2006)
20. Y. Tsukamoto, K. Nii, S. Imaoka, Y. Oda, S. Ohbayashi, T. Yoshizawa, H. Makino, K. Ishibashi, H. Shinohara, Worst-case analysis to obtain stable read/write DC margin of high density 6T-SRAM-array with local Vth variability, in *IEEE/ACM International Conference on Computer-Aided Design, ICCAD-2005*, pp. 398–405, Nov 2005
21. H. Mahmoodi, S. Mukhopadhyay, K. Roy, Estimation of delay variations due to random-dopant fluctuations in nanoscale CMOS circuits. IEEE J. Solid-State Circuits **40**(9), 1787–1796 (2005)
22. R. Kanj, R. Joshi, S. Nassif, Mixture importance sampling and its application to the analysis of SRAM designs in the presence of rare failure events, in *43rd ACM/IEEE Design Automation Conference*, 0–0 2006, pp. 69–72
23. T. Date, S. Hagiwara, K. Masu, T. Sato, Robust importance sampling for efficient SRAM yield analysis, in *11th International Symposium on Quality Electronic Design (ISQED)*, pp. 15–21, March 2010
24. C. Dong, X. Li, Efficient SRAM failure rate prediction via Gibbs sampling, in *Proceedings of the 48th Design Automation Conference, ser. DAC '11* (ACM, New York, 2011), pp. 200–205
25. L. Dolecek, M. Qazi, D. Shah, A. Chandrakasan, Breaking the simulation barrier: SRAM evaluation through norm minimization, in *IEEE/ACM International Conference on Computer-Aided Design, ICCAD* 2008, pp. 322–329, Nov 2008
26. D. Khalil, M. Khellah, N.-S. Kim, Y. Ismail, T. Karnik, V. De, Accurate estimation of SRAM dynamic stability. IEEE Trans. Very Large Scale Integr. (VLSI) Syst. **16**(12), 1639–1647 (2008)
27. A. Singhee, R. Rutenbar, Statistical blockade: very fast statistical simulation and modeling of rare circuit events and its application to memory design. IEEE Trans. Comput. Aided Des. Integr. Circuits Syst. **28**(8), 1176–1189 (2009)
28. R. Aitken, S. Idgunji, Worst-case design and margin for embedded SRAM, in *Design, Automation Test in Europe Conference Exhibition, DATE '07*, pp. 1–6, April 2007
29. J. Wang, A. Singhee, R. Rutenbar, B. Calhoun, Statistical modeling for the minimum standby supply voltage of a full SRAM array, in *Proceedings of the 33rd European Solid State Circuits Conference, ESSCIRC*, pp. 400–403, Sept 2007
30. J. Wang, A. Singhee, R. Rutenbar, B. Calhoun, Two fast methods for estimating the minimum standby supply voltage for large SRAMs. IEEE Trans. Comput. Aided Des. Integr. Circuits Syst. **29**(12), 1908–1920 (2010)

31. A. Singhee, J. Wang, B. Calhoun, R. Rutenbar, Recursive statistical blockade: an enhanced technique for rare event simulation with application to sram circuit design, in *Proceedings of the 21st International Conference on VLSI Design, VLSID 2008*, pp. 131–136, Jan 2008
32. B. Amrutur, M. Horowitz, Speed and power scaling of SRAM's. IEEE J. Solid-State Circuits **35**(2), 175–185 (Feb 2000)
33. M. Yamaoka, T. Kawahara, Operating-margin-improved SRAM with column-at-a-time body-bias control technique, in *Proceedings of the 33rd European Solid State Circuits Conference, ESSCIRC*, pp. 396–399, 11–13 Sept 2007
34. H. Pilo, IEDM SRAM short course, 2006
35. M. Pelgrom, H. Tuinhout, M. Vertregt, Transistor matching in analog CMOS applications, in *Proceedings of the International Electron Devices Meeting (IEDM)*, 1998, pp. 915–918
36. Y. Taur, T.H. Ning, *Fundamentals of Modern VLSI Devices* (Cambridge University Press, Cambridge, 1998)
37. Y. Wang, H.J. Ahn, U. Bhattacharya, Z. Chen, T. Coan, F. Hamzaoglu, W. Hafez, C.-H. Jan, P. Kolar, S. Kulkarni, J.-F. Lin, Y.-G. Ng, I. Post, L. Wei, Y. Zhang, K. Zhang, M. Bohr, A 1.1 GHz 12 A/Mb-Leakage SRAM design in 65 nm ultra-low-power CMOS technology with integrated leakage reduction for mobile applications. IEEE J. Solid-State Circuits **43**(1), 172–179 (Jan. 2008)
38. B. Wicht, T. Nirschl, D. Schmitt-Landsiedel, Yield and speed optimization of a latch-type voltage sense amplifier. IEEE J. Solid-State Circuits **39**(7), 1148–1158 (July 2004)
39. T. Matthews, P. Heedley, A simulation method for accurately determining dc and dynamic offsets in comparators, in *Proceedings of the 48th Midwest Symposium on Circuits and Systems*, vol. 2, pp. 1815–1818, Aug 2005
40. P. Kinget, Device mismatch and tradeoffs in the design of analog circuits. IEEE J. Solid-State Circuits **40**(6), 1212–1224 (2005)
41. S. Mukhopadhyay, K. Kim, K. Jenkins, C.-T. Chuang, K. Roy, Statistical characterization and on-chip measurement methods for local random variability of a process using sense-amplifier-based test structure, in ltextitProceedings of the International Solid-State Circuits Conference ISSCC, pp. 400–611, Feb 2007
42. A Methodology for the Offset-Simulation of Comparators, http://www.designers-guide.org/Analysis/comparator.pdf
43. T. Matthews, P. Heedley, A simulation method for accurately determining dc and dynamic offsets in comparators, in *Proceedings of the 48th Midwest Symposium on Circuits and Systems*, vol. 2, pp. 1815–1818, Aug 2005
44. B. Razavi, B.A. Wooley, Design techniques for high-speed, high-resolution comparators. IEEE J. Solid-State Circuits **27**, 1916–1926 (1992)
45. J. Ryan, B. Calhoun, Minimizing offset for latching voltage-mode sense amplifiers for sub-threshold operation, in *Proceedings of the 9th International Symposium on Quality Electronic Design, ISQED* 2008, pp. 127–132, March 2008
46. R. Singh, N. Bhat, An offset compensation technique for latch type sense amplifiers in high-speed low-power SRAMs. IEEE Trans. Very Large Scale Integr. (VLSI) Syst. **12**(6), 652–657 (2004)
47. D.A. Schinkel, E. Mensink, E. Klumperink, E.T. van, A double-tail latch-type voltage sense amplifier with 18ps setup+hold time, in *Proceedings of the International Solid-State Circuits Conference ISSCC*, pp. 314–315, 2007
48. A. Nikoozadeh, B. Murmann, An analysis of latch comparator offset due to load capacitor mismatch. IEEE Trans. Circuits Syst. II: Express Briefs **53**(12), 1398–1402 (2006)
49. B. Razavi, *Design of Analog CMOS Integrated Circuits* (McGraw-Hill, New York, 2000)
50. B. Leibowitz, J. Kim, J. Ren, C. Madden, Characterization of random decision errors in clocked comparators, in *IEEE Custom Integrated Circuits Conference, CICC*, pp. 691–694, Sept 2008
51. B. Amrutur, M. Horowitz, A replica technique for wordline and sense control in low-power SRAM's. IEEE J. Solid-State Circuits **33**(8), 1208–1219 (Aug 1998)
52. K. Bowman, S. Duvall, J. Meindl, Impact of die-to-die and within-die parameter fluctuations on the maximum clock frequency distribution for gigascale integration. IEEE J. Solid-State Circuits **37**(2), 183–190 (2002)

53. E. Morifuji, D. Patil, M. Horowitz, Y. Nishi, Power optimization for SRAM and its scaling. IEEE Trans. Electron Devices **54**(4), 715–722 (April 2007)
54. C. Pacha, B. Martin, K. von Arnim, R. Brederlow, D. Schmitt-Landsiedel, P. Seegebrecht, J. Berthold, R. Thewes, Impact of STI-induced stress, inverse narrow width effect, and statistical v_{TH} variations on leakage currents in 120 nm CMOS, in *Proceeding of the 34th European Solid-State Device Research conference ESSDERC*, pp. 397–400, 2004

Chapter 6
Characterization of SRAM Sense Amplifier Input Offset for Yield Prediction

6.1 Introduction

In nanometer technologies, embedded SRAM occupies a significant portion of system on chips (SoCs) and has a large impact on chip yield. With the increase in random variation with scaling (due to RDF, LER, and other sources), SRAMs become extremely sensitive to process variations, especially as the supply voltage is reduced. Random variations reduce yield due to several mechanisms, such as read stability, writeability, retention, and read sense margin (access yield); however, read sense margin is the mechanism that typically limits SRAM speed [1–3].

Measurement of device characteristics and variability is critical for SRAM optimization. Instead of measuring basic device parameters using simple test structures, it is more useful to directly measure the quantity of interest and apply them to the design [4]. For yield sensitive blocks such as SRAM, it is therefore important to develop process monitors to measure the SA input offset and use measurement results to improve memory yield and improve access time. Moreover, silicon validation is essential to avoid excessive conservatism in design margining techniques and achieve competitive designs. Several methods to measure SRAM parameters using complex test structures have been proposed [3, 5–15]. These structures are important for technology development and are extensively used to evaluate different read and write assist techniques, impact of aging and noise on SRAM stability. Figure 6.1 shows a test structure used to measure SRAM read current and stability using ring oscillators [8].

Sense amplifiers (SA) are used extensively in memory read operation to amplify the small signal bitline voltage differential to the digital level. SAs suffer from random variations, which increase the SA input offset (due to its differential nature and small signal operation) and thus decreases SRAM read access yield (Y_{read}) [2, 16, 17]. In this chapter, we focus on Y_{read} instead of other stability failures, due to its strong sensitivity to SA offset and its impact on SRAM performance (i.e., access time).

SA input offset is typically estimated by Monte Carlo simulation and varying device mismatch parameters. Mismatch parameters are usually extracted using simple device pair structures with large distances between neighboring structures [18].

M. H. Abu-Rahma and M. Anis, *Nanometer Variation-Tolerant SRAM*,
DOI: 10.1007/978-1-4614-1749-1_6, © Springer Science+Business Media New York 2013

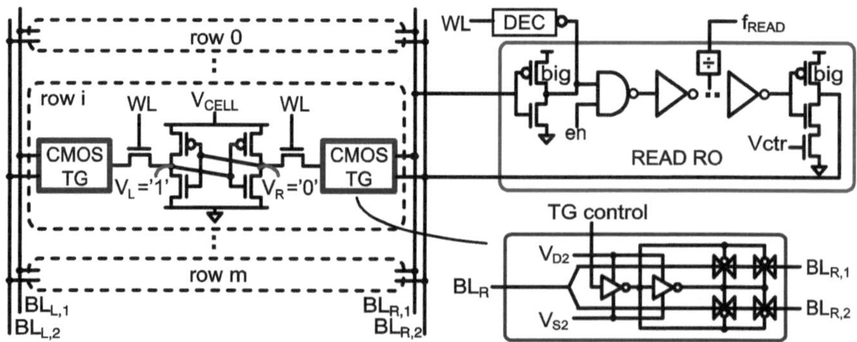

Fig. 6.1 Ring oscillator test structure used to measure read stability and bitcell read current [8]

This modeling technique can be adequate for designs that are not area constrained since proximity effects can be reduced by increasing design area. However, in SRAM design, the memory area is very constrained, and proximity effects can play a major role in determining SA offset. Proximity effects (e.g., well proximity, stress effects, and active area rounding) can impact the accuracy of SA offset estimation, either due to inadequate modeling for physical effects, or due to limited understanding of its impact on mismatch. These modeling limitations can introduce large errors in the estimation of SA offset during design, which decrease the memory performance and yield.

An example of mismatch modeling limitations is shown in Fig. 6.2 where silicon measurements deviate significantly from standard RDF theory [19, 20]. Silicon data show that the relationship between $\sigma_{V\text{th}}$ and $1/\text{sqrt}(WL)$ is not linear as predicted by theory [21], but instead the mismatch slope $A_{V\text{th}}$ depends strongly on channel length L, as shown in Fig. 6.2 [19]. Work on modeling device variability using measurements and TCAD has shown that halo dopants dominate V_{th} variation, especially for longer channel devices [19, 22–24]. This effect as well as other second-order effects cannot be modeled using the simple mismatch models [21]. In SRAM sense amplifiers, longer channel length devices are typically used to reduce the offset, which highlights the importance of $\sigma_{V_{\text{th}}}$ and offset measurements.

6.2 SRAM Read Access Yield Sensitivity to SA Offset

SRAM read access yield (Y_{read}) is defined as the probability of correct read sensing operation [1, 2, 16, 17]. Y_{read} depends on SA input offset, bitcell read current variations (I_{read}), variation in the delay between wordline enable to SA activation, and bitcell pass-gate leakage. With the increase in variations with scaling, recently, accurate characterization of SRAM bitcell has been gaining much attention and methods to directly measure bitcell variations have been proposed [11, 25, 26]. To achieve

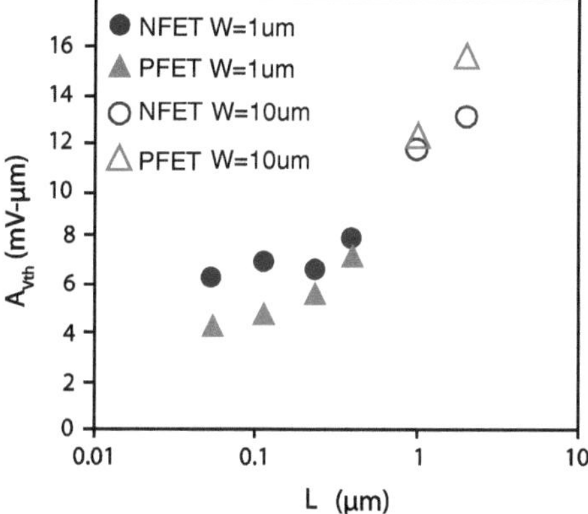

Fig. 6.2 Mismatch coefficient $A_{V_{th}}$ versus L for 65 nm low-power technology, which the strong dependence of $A_{V_{th}}$ on L. This trend cannot be explained by classical RDF theory which predicts that $A_{V_{th}}$ shows no dependence on L [19]

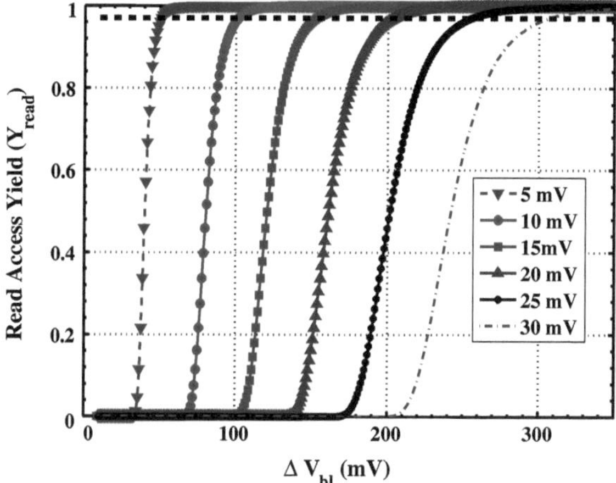

Fig. 6.3 Statistical yield simulation for a 16 Mb memory in 28 nm showing Y_{read} versus ΔV_{bl} for different values of SA input offset σ_{offset}. The horizontal dotted line marks the 97 % yield target. As σ_{offset} increases, a larger ΔV_{bl} is needed to achieve the same yield target

higher memory speed, it is important to reduce the bitline differential ΔV_{bl} sensed by the SA, while ensuring that the yield target is met. Accurate characterization of the SA input offset is key to achieve this goal.

To evaluate the impact of SA offset on Y_{read}, we simulated the read access yield for a 16 Mb memory in a 28 nm LP technology node using statistical Monte-Carlo yield estimation techniques that account for I_{read} and SA offset variations [2], as presented in Chap. 5. Figure 6.3 shows the relationship between Y_{read} and the mean ΔV_{bl} for different values of SA offset. The increase in SA input offset (σ_{offset}) decreases Y_{read}.

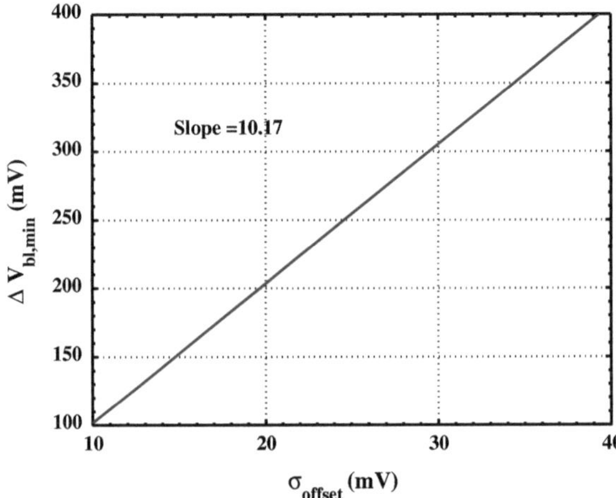

Fig. 6.4 Minimum required V_{bl} versus σ_{offset} at a constant yield target (97 % for 16 Mb)

To achieve higher yield, ΔV_{bl} should be increased, which reduces the SRAM speed, as the bitcell development time (time to generate ΔV_{bl}) is typically 30–50 % of the memory access time. Therefore, it is of utmost importance to design for the minimum ΔV_{bl} ($\Delta V_{\text{bl,min}}$) while meeting the yield target. Accurate measurement of the σ_{offset} allows the designer to set $\Delta V_{\text{bl,min}}$ without loss in yield or performance.

Figure 6.4 shows the significant increase in $\Delta V_{\text{bl,min}}$ as σ_{offset} is increased for a high yield target (97 % for 16 Mb memory). For every 1 mV increase in offset, the required $\Delta V_{\text{bl,min}}$ increases by 10 mV, which decreases speed, since the bitline discharge time is usually large due to the large bitline capacitance. Moreover, increasing the SA area to reduce its input offset increases the bitline capacitance (reduces speed) and increases the memory area. This tradeoff between memory performance and SA input offset highlights the importance of accurate silicon characterization of σ_{offset} for SRAM yield prediction, especially in the early process technology development phases.

6.3 Sense Amplifier Offset Monitor Implementation

Recently, an SA-based process monitor has been proposed to measure random V_{th} variations [27]. In this work, we extend the idea to measure σ_{offset} by utilizing SAs used in 28 nm memories. The proposed monitor employs two type of SAs widely used in SRAM design: (1) voltage latch sense amplifier (VLSA) [28] and (2) current latch sense amplifier (CLSA) [29], as shown in Fig. 6.6. Since memories are typically area constrained, we implemented the two SAs topologies using the same total area, which allows the offset values to be compared while taking memory design constraints into account.

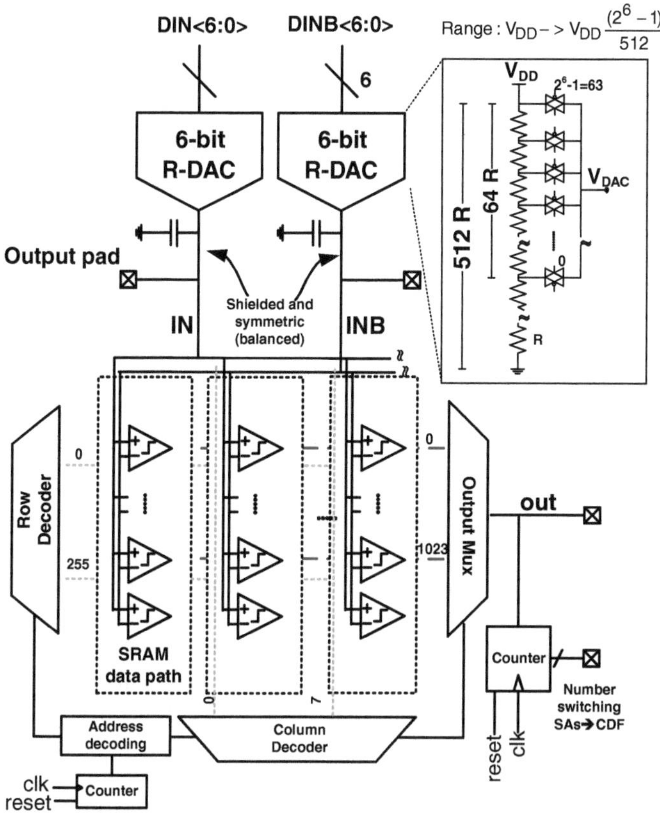

Fig. 6.5 Implementation of the SA input offset monitor

In nanometer technologies, layout proximity effects (e.g., well-proximity, stress effects, critical dimensions rounding) strongly affect device characteristics. To account for proximity effects around the SA, the test structure uses the same SA layout used in 28 nm memory macros, which are pitch matched to 4 bitcell columns. Moreover, the SA layout includes all the associated read/write circuitry around the SA in a memory data path (e.g., write drivers, precharge devices, bitline muxes). Therefore, the measured offset accounts for all proximity effects and provides the accurate estimate needed for yield calculation.

The proposed monitor consists of arrays of 1024 addressable SAs, and SA inputs are driven using a two 6-bit digital to analog converters (DAC) as shown in Fig. 6.5. A resistor string DAC is chosen for its excellent linearity, immunity against die-to-die variations, low voltage operation, and ease of migration to newer technologies. In addition, the on-chip DAC eliminates the complexity of driving the SA inputs from an external voltage source, which helps reduce the test time.

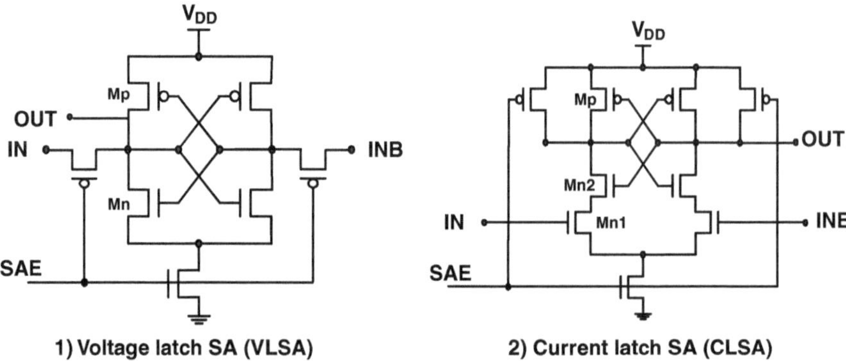

Fig. 6.6 Sense amplifiers used in this study: **a** voltage latch sense amplifier (VLSA) [28] **b** current latch sense amplifier (CLSA) [29]

The DAC allows the SA input to vary within the range $\pm V_{\mathrm{DD}} * (2^6 - 1)/512$ in a step of $V_{\mathrm{DD}}/512$, as shown in Fig. 6.5. The DAC transmission gates are sized appropriately to reduce the voltage drop on the gates due to SA input leakage (gate tunneling for CLSA, subthreshold/junction/gate leakage for VLSA). The output from the DAC is shielded to reduce coupling noise, and is distributed symmetrically to all the SAs to eliminate any systematic offset. To reduce kickback noise on the SA inputs, large capacitors are added on the DAC output and only one SA is selected at a time (other SAs are clock gated). The output from all the SAs is muxed to the chip output, and to a digital counter that counts the number of SAs toggling, from which the distribution of SA offset can be extracted. The output of the DAC is also monitored off-chip to characterize the DAC performance and compensate or calibrate as needed.

The SA offset distribution is measured by setting the DAC input code, and then testing all the addressable SAs sequentially. The number of SAs that switch correctly at a specific DAC setting is recorded using the output counter. The SA yield can be calculated as the ratio of the number of SA that switched correctly to the total number of SAs (1024). The same test operation is repeated for the whole range of the DAC to extract the SA input offset cumulative distribution function (CDF).

6.4 Results and Discussion

The proposed offset monitor was implemented in a test chip in a 28 nm LP process [30]. In addition, a 512 Kb memory was implemented in the same test chip to characterize Y_{read} versus $\Delta V_{\mathrm{bl,min}}$. Table 6.1 shows the sizing details of the SAs.

Measured DAC results are shown in Fig. 6.7. DAC output was measured on 30 dies and the results show negligible die-to-die spread. The DAC integral nonlinearity (INL) and differential nonlinearity (DNL) is very small, as shown in Fig. 6.8a and Fig. 6.8b, respectively, and both INL and DNL have tight distributions as indicated

Table 6.1 Sizing details and the measured offset

Type	VLSA	CLSA [b]
Wn (nm) [a]	835 × 4	835 × 2
L(nm)	55	55
σ_{offset} (mV) [c]	9.5	26.7
CI (mV) [d]	0.27	0.65
$S_{V_{\text{DD}}}$ (%/%) [e]	−0.55	0.25
S_T (%/°C) [f]	−0.12	−0.11

[a] All sizes are drawn dimensions. [b] Mn1 and Mn2 shown in Fig. 6.6 have the same size. [c] Measured at nominal V_{DD} and 25 °C. [d] 95 % statistical confidence. [e] $S_{V_{\text{DD}}}$: V_{DD} sensitivity. [f] S_T: Temperature sensitivity

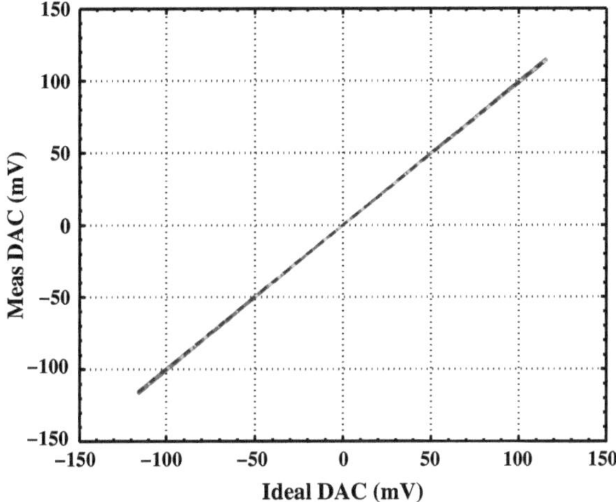

Fig. 6.7 Measured DAC versus ideal DAC characteristics for 30 dies. Each die is a *different color line*; *lines* are superimposed

by the linearity variation (σ_{INL} and σ_{DNL}). These measurements confirm that the DAC does not introduce errors in the measurement of the SA offset.

The all-digital nature of the monitor allows collecting large number of statistical samples in a short test time. The offset monitor operates at high speed (10 MHz) compared to conventional device mismatch characterization. The measured offset distribution for ~35K SAs is shown in Fig. 6.9. The distributions are Gaussian within a range of $\pm 3.5\sigma$ away from the mean. For the same area, the VLSA offset distribution is much tighter than CLSA; the measured σ_{offset} is 9.5 and 26.7 mV, for VLSA and CLSA, respectively, the confidence interval of the offset distributions is very small due to the large sample size used, indicating high accuracy of the offset measurements (Table 6.1).

Detailed offset characterization was performed for different temperature and voltage conditions. Figure 6.10 shows the dependence of offset on temperature; both VLSA and CLSA follow similar trends, where offset decreases at higher temperature.

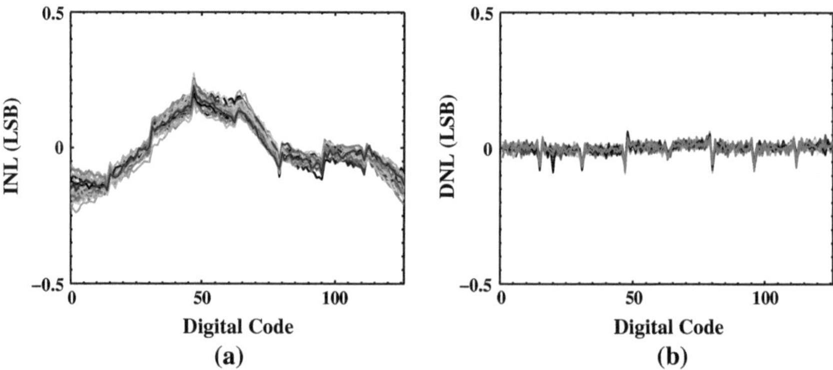

Fig. 6.8 DAC linearity characterization for 30 dies; each die is a different color. **a** Measured integral nonlinearity (INL)($\mu_{\mathrm{INL}} = 0.223$ LSB and $\sigma_{\mathrm{INL}} = 0.027$ LSB) **b** Measured differential nonlinearity (DNL) $\mu_{\mathrm{DNL}} = 0.076$ LSB and $\sigma_{\mathrm{DNL}} = 0.027$ LSB

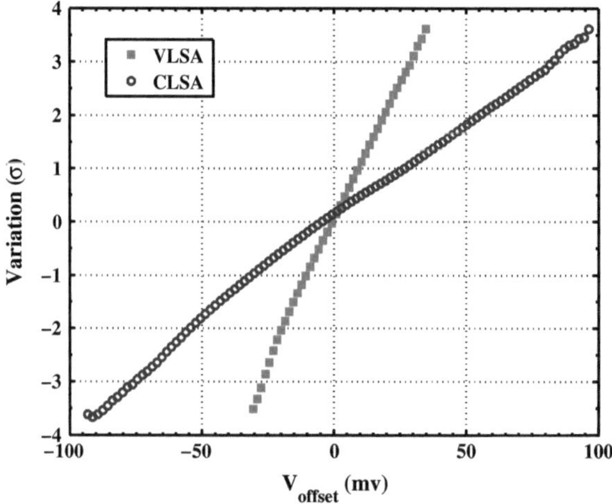

Fig. 6.9 Measured SA offset cumulative distribution for 35 k sense amplifiers (CLSA and VLSA types of same area). The distributions follow a Gaussian shape $\pm 3.5\sigma$ away from the mean

This trend may be explained by recent findings that device mismatch decreases as temperature increases [31]. Figure 6.11 shows the temperature dependence on mismatch parameters ($\sigma_{V_{\mathrm{th}}}$), the current factor $\sigma_{\Delta\beta/\beta}$ and $\sigma_{\Delta I_{\mathrm{ON}}/I_{\mathrm{ON}}}$) for NMOS and PMOS devices in a 65 nm LP CMOS technology [31]. Measurements show that $\sigma_{V_{\mathrm{th}}}$ slightly improves at higher temperature, while $\sigma_{\Delta\beta/\beta}$ shows significant reduction. β is proportional to W/L ratio, gate oxide capacitance C_{ox}, and the mobility μ. It is well-known that μ depends strongly on temperature [32]. At higher temperature, μ is dominated by the lattice scattering component and not dopants scattering which is the major contributor for β mismatch. The improvement in $\sigma_{V_{\mathrm{th}}}$ and β in addition to the reduction in V_{th} at higher temperature help to improve I_{ON} mismatch at

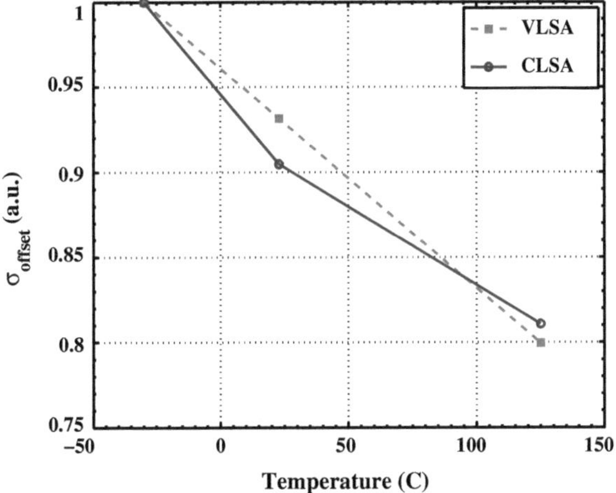

Fig. 6.10 Measured σ_{offset} versus temperature for VLSA and CLSA

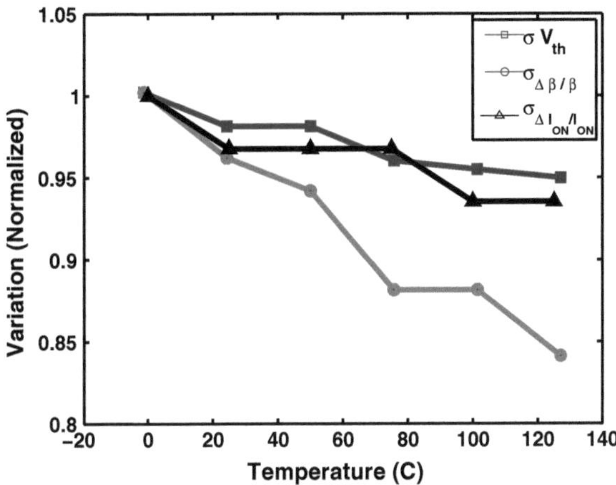

Fig. 6.11 Measured variation $\sigma_{V_{\text{th}}}$, $\sigma_{\Delta\beta/\beta}$, and $\sigma_{\Delta I_{\text{ON}}/I_{\text{ON}}}$ versus temperature for 65 nm NMOS which shows reduction of variation at higher temperature [31]

higher temperature. Therefore, this trend can help explain the reduction of SA offset variation with temperature as shown in Fig. 6.10.

Low-voltage operation of memories is important for low-power designs. It is well-known that bitcell I_{read} variation (σ/μ) increases significantly as V_{DD} is lowered. The impact of V_{DD} reduction on offset is shown in Fig. 6.12. Interestingly, VLSA and CLSA show opposite sensitivities to V_{DD}. The VLSA σ_{offset} increases as V_{DD} is lowered, while CLSA offset decreases. This may suggest that CLSA is better suited for low-voltage operation. Nevertheless, the VLSA area efficiency and lower offset makes it more attractive even though its V_{DD} sensitivity is negative. These trends

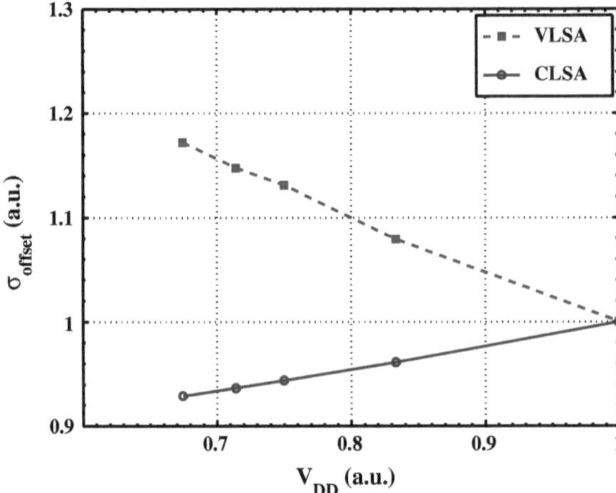

Fig. 6.12 Measured σ_{offset} versus supply voltage V_{DD} for VLSA and CLSA

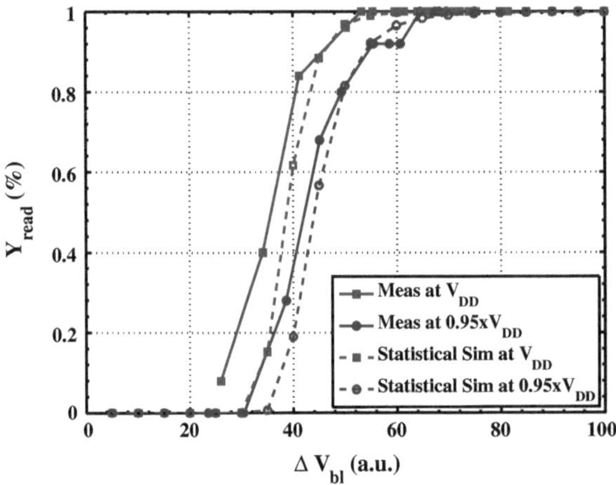

Fig. 6.13 Measured and simulated Y_{read} versus ΔV_{bl} for a 512 Kb memory. The statistical simulation includes the measured silicon offset

show the importance of accurate silicon characterization of offset for SRAM yield estimation. Table 6.1 shows a summary of the measured characteristics of the two SA types.

Using the silicon measured offset from the proposed test structure and the distribution of bitcell read current (from a separate test circuit [26]), statistical yield simulation was performed to correlate the measured Y_{read} to the simulation. As shown in Fig. 6.13, statistical simulation agrees with silicon for different supply voltages. This correlation would not be possible without the accurately measured SA offset.

This illustrates the importance of Si-accurate SA offset in SRAM yield prediction and performance optimization.

6.5 Summary

An SA offset monitor was proposed and implemented in 28 nm LP CMOS technology to predict SRAM yield. The monitor accurately characterizes for the SA input offset at different conditions. Silicon data shows that high temperature decreases input offset in both VLSA and CLSA type SAs. VLSA offset increases at lower V_{DD} while CLSA shows the opposite trend. These trends show the importance of accurate offset measurement for SRAM yield optimization and design/process improvement.

The application of the Si-accurate offset in correlating measured memory yield to statistical simulation was established using a 512 Kb memory implemented on the same test chip. The monitor as well as other SRAM test structures will likely become a critical part of SRAM silicon yield validation, which is becoming more and more important with technology scaling and the increase in random variations.

References

1. S. Mukhopadhyay, H. Mahmoodi, K. Roy, Statistical design and optimization of SRAM cell for yield enhancement, in *Proceedings of International Conference on, Computer Aided Design*, 2004, pp. 10–13
2. M. H. Abu-Rahma, K. Chowdhury, J. Wang, Z. Chen, S. S. Yoon, M. Anis, A methodology for statistical estimation of read access yield in SRAMs, in *Proceedings of the 45th Conference on Design Automation*, 2008, pp. 205–210
3. M. Abu-Rahma, Y. Chen, W. Sy, W. L. Ong, L. Y. Ting, S. S. Yoon, M. Han, E. Terzioglu, "Characterization of SRAM sense amplifier input offset for yield prediction in 28 nm CMOS", in *IEEE Custom Integrated Circuits Conference (CICC)*, September 2011
4. F. Klass, A. Jain, G. Hess, B. Park, An all-digital on-chip process-control monitor for process-variability measurements, in *Proceedings of the International Solid-State Circuits Conference (ISSCC)*, 2008, pp. 408–623
5. R. Venkatraman, R. Castagnetti, S. Ramesh, The statistics of device variations and its impact on SRAM bitcell performance, leakage and stability, in *Proceedings of the International Symposium on Quality of Electronic Design (ISQED)*, 2006, pp. 190–195
6. S. O. Toh, Z. Guo, and B. Nikolić and, "Dynamic SRAM stability characterization in 45 nm CMOS", in *IEEE Symposium on VLSI Circuits (VLSIC)*, June 2010, pp. 35–36
7. Z. Guo, A. Carlson, L.-T. Pang, K. Duong, T.-J. K. Liu, B. Nikolić, Large-scale SRAM variability characterization in 45 nm CMOS, IEEE J. Solid-State Circuits, **44**(11), pp. 3174–3192, November 2009
8. J. Tsai, S.O. Toh, Z. Guo, L.-T. Pang, T.-J.K. Liu, B. Nikolić, SRAM stability characterization using tunable ring oscillators in 45nm CMOS, in *IEEE. International Solid-State Circuits Conference Digest of Technical Papers (ISSCC)*, February 2010, 354–355
9. M. Qazi, K. Stawiasz, L. Chang, A. Chandrakasan, A 512kb 8T SRAM macro operating down to 0.57 v with an AC-coupled sense amplifier and embedded data-retention-voltage sensor in 45 nm SOI CMOS, in *IEEE. International Solid-State Circuits Conference Digest of Technical Papers (ISSCC)*, February 2010, 350–351

10. B. Nikolić and, B. Giraud, Z. Guo, L.-T. Pang, J.-H. Park, and S. O. Toh, "Technology variability from a design perspective", in *IEEE Custom Integrated Circuits Conference (CICC)*, September 2010, pp. 1–8
11. T. Fischer, C. Otte, D. Schmitt-Landsiedel, E. Amirante, A. Olbrich, P. Huber, M. Ostermayr, T. Nirschl, J. Einfeld, "A 1 Mbit SRAM test structure to analyze local mismatch beyond 5 sigma variation", in *Proceedings of the IEEE International Conference on Microelectronic Test Structures (ICMTS)*, pp. 63–66, 2007
12. B. Nikolic, J.-H. Park, J. Kwak, B. Giraud, Z. Guo, L.-T. Pang, S. O. Toh, R. Jevtic, K. Qian, C. Spanos, "Technology variability from a design perspective", In: IEEE Trans.Regul. Pap.Circ and Syst.,**58**(9), pp. 1996–2009, September 2011
13. Y.-H. Chen, S.-Y. Chou, Q. Lee, W.-M. Chan, D. Sun, H.-J. Liao, P. Wang, M.-F. Chang, H. Yamauchi, A 40 nm fully functional SRAM with bl swing and wl pulse measurement scheme for eliminating a need for additional sensing tolerance margins, in *Symposium on VLSI Circuits (VLSIC)*, June 2011, pp. 70–71
14. Y.-H. Chen, S.-Y. Chou, Q. Li, W.-M. Chan, D. Sun, H.-J. Liao, P. Wang, M.-F. Chang, H. Yamauchi, "Compact measurement schemes for bit-line swing, sense amplifier offset voltage, and word-line pulse width to characterize sensing tolerance margin in a 40 nm fully functional embedded SRAM", IEEE J. Solid-State Circ.**PP**(99) 1–12 (2012)
15. M. Suzuki, T. Saraya, K. Shimizu, A. Nishida, S. Kamohara, K. Takeuchi, S. Miyano, T. Sakurai, and T. Hiramoto, "Direct measurements, analysis, and post-fabrication improvement of noise margins in SRAM cells utilizing DMA SRAM TEG", in *Symposium on VLSI Technology (VLSIT)*, June 2010, pp. 191–192
16. R. Heald and P. Wang, "Variability in sub-100 nm SRAM designs", in *Proceedings of International conference on, Computer Aided Design (ICCAD)*, 2004, pp. 347–352
17. R. Houle, Simple statistical analysis techniques to determine minimum sense amp set times,*Proceedings of IEEE Custom Integrated Circuits conference*, pp. 37–40, 2007
18. U. Schaper, J. Einfeld, A. Sauerbrey, "Parameter Variation on Chip-Level", in *Proceedings of the 2005 International Conference on Microelectronic Test Structures (ICMTS)*, 2005
19. J. Johnson, T. Hook, Y.-M. Lee, "Analysis and modeling of threshold voltage mismatch for CMOS at 65 nm and beyond", IEEE Electron Device Lett. **29**(7), pp. 802–804
20. J. Mc Ginley, O. Noblanc, C. Julien, S. Parihar, K. Rochereau, R. Difrenza, P. Llinares, Impact of pocket implant on MOSFET mismatch for advanced CMOS technology, in *Proceedings of The International Conference on Microelectronic Test Structures (ICMTS)*, March 2004, 123–126
21. M. Pelgrom, A. Duinmaijer, A. Welbers, Matching properties of MOS transistors. IEEE J. Solid-State Circ. **24**(5), 1433–1439 (1989)
22. C.M. Mezzomo, A. Bajolet, A. Cathignol, E. Josse, G. Ghibaudo, Modeling local electrical fluctuations in 45 nm heavily pocket-implanted bulk mosfet, SolidState Electron.**54**(11), pp. 1359–1366 (2010)
23. C. M. Mezzomo, A. Bajolet, A. Cathignol, G. Ghibaudo, Drain-current variability in 45 nm bulk n-mosfet with and without pocket-implants, Solid-State Electronics, **26**(3), 1920-1926 (2011)
24. C. Mezzomo, A. Bajolet, A. Cathignol, R. Di Frenza, G. Ghibaudo, Characterization and modeling of transistor variability in advanced CMOS technologies, IEEE Trans. Electron. Devices, **58**, (8), pp. 2235–2248, August 2011
25. X. Deng, W. K. Loh, B. Pious, T. Houston, L. Liu, B. Khan, D. Corum, Characterization of bit transistors in a functional SRAM, *Proceedings of IEEE Symposium on VLSI Circuits*, pp. 44–45, 2008
26. J. Wang, P. Liu, Y. Gao, P. Deshmukh, S. Yang, Y. Chen, W. Sy, L. Ge, E. Terzioglu, M. Abu-Rahma, M. Garg, S. S. Yoon, M. Han, M. Sani, and G. Yeap, Non-Gaussian distribution of SRAM read current and design impact to low power memory using voltage acceleration method, in *Symposium on VLSI Technology (VLSIT)*, June 2011, pp. 220–221

27. S. Mukhopadhyay, K. Kim, K. Jenkins, C.-T. Chuang, K. Roy, Statistical characterization and on-chip measurement methods for local random variability of a process using sense-amplifier-based test structure, in *Proceedings of the International Solid-State Circuits Conference (ISSCC)*, pp. 400–611, Feb. 2007
28. B. Razavi, B.A. Wooley, Design techniques for high-speed, high-resolution comparators. IEEE J. SolidState Circ. **27**, 1916–1926 (1992)
29. T. Kobayashi, K. Nogami, T. Shirotori, Y. Fujimoto, A current-controlled latch sense amplifier and a static power-saving input buffer for low-power architecture, IEEE J. SolidState Circ. **28**(4), 523–527 (1993)
30. P. Chidambaram, C. Gan, S. Sengupta, L. Ge, Y. Chen, S. Yang, P. Liu, J. Wang, M. Yang, C. Teng, Y. Du, P. Patel, P. Kamal, R. Bucki, F. Vang, A. Datta, K. Bellur, S. Yoon, N. Chen, A. Thean, M. Han, E. Terzioglu, X. Zhang, J. Fischer, M. Sani, B. Flederbach, G. Yeap, Cost effective 28 nm LP SoC technology optimized with circuit/device/process co-design for smart mobile devices, in *IEEE, International Electron Devices Meeting (IEDM)*, December 2010, pp. 27.3.1–27.3.4
31. P. Andricciola, H. Tuinhout, The temperature dependence of mismatch in deep-submicrometer bulk MOSFETs. IEEE Electron Device Lett. **30**(6), 690–692 2009
32. Y. Cheng and C. Hu, MOSFET Modeling and BSIM User Guide. (Kluwer Academic Publishers, Massachusetts, 1999)

Index

M. H. Abu-Rahma and M. Anis, *Nanometer Variation-Tolerant SRAM*,
DOI: 10.1007/978-1-4614-1749-1, © Springer Science+Business Media New York 2013

If you have any concerns about our products,
you can contact us on
ProductSafety@springernature.com

In case Publisher is established outside the EU,
the EU authorized representative is:
Springer Nature Customer Service Center GmbH
Europaplatz 3, 69115 Heidelberg, Germany

Printed by Libri Plureos GmbH
in Hamburg, Germany